THE HISTORY *of* MEDICINE

OLD WORLD AND NEW

EARLY MEDICAL CARE, 1700–1840

THE HISTORY *of* MEDICINE

OLD WORLD AND NEW

EARLY MEDICAL CARE, 1700–1840

KATE KELLY

Facts On File
An imprint of Infobase Publishing

OLD WORLD AND NEW: Early Medical Care, 1700–1840

Facts On File, Inc.
An imprint of Infobase Publishing
132 West 31st Street
New York NY 10001

Library of Congress Cataloging-in-Publication Data
Kelly, Kate.
    Old world and new: early medical care, 1700–1840 / Kate Kelly.
        p. cm. — (The history of medicine)
    Includes bibliographical references and index.
    ISBN 978-0-8160-7208-8
    1. Medicine—History—18th century. 2. Medicine—United States—History—18th century. I. Title.

R148.K45 2010
610—dc22   2009005163

Facts On File books are available at special discounts when purchased in bulk quantities for businesses, associations, institutions, or sales promotions. Please call our Special Sales Department in New York at (212) 967-8800 or (800) 322-8755.

You can find Facts On File on the World Wide Web at http://www.factsonfile.com

Excerpts included herewith have been reprinted by permission of the copyright holders; the author has made every effort to contact copyright holders. The publishers will be glad to rectify, in future editions, any errors or omissions brought to their notice.

Text design by Annie O'Donnell
Illustrations by Bobbi McCutcheon
Photo research by Elizabeth H. Oakes
Composition by Hermitage Publishing Services
Cover printed by Bang Printing, Inc., Brainerd, Minn.
Book printed and bound by Bang Printing, Inc., Brainerd, Minn.
Date printed: November, 2009
Printed in the United States of America

10 9 8 7 6 5 4 3 2 1

This book is printed on acid-free paper.

CONTENTS

"You have to know the past to understand the present."
—American scientist Carl Sagan (1934–96)

The history of medicine offers a fascinating lens through which to view humankind. Maintaining good health, overcoming disease, and caring for wounds and broken bones was as important to primitive people as it is to us today, and every civilization participated in efforts to keep its population healthy. As scientists continue to study the past, they are finding more and more information about how early civilizations coped with health problems, and they are gaining greater understanding of how health practitioners in earlier times made their discoveries. This information contributes to our understanding today of the science of medicine and healing.

In many ways, medicine is a very young science. Until the mid-19th century, no one knew of the existence of germs, so as a result, any solutions that healers might have tried could not address the root cause of many illnesses. Yet for several thousand years, medicine has been practiced, often quite successfully. While progress in any field is never linear (very early, nothing was written down; later, it may have been written down, but there was little intra-community communication), readers will see that some civilizations made great advances in certain health-related areas only to see the knowledge forgotten or ignored after the civilization faded. Two early examples of this are Hippocrates' patient-centered healing philosophy and the amazing contributions of the Romans to public health through water-delivery and waste-removal systems. This knowledge was lost and had to be regained later.

The six volumes in the History of Medicine set are written to stand alone, but combined, the set presents the entire sweep of the history of medicine. It is written to put into perspective for high school students and the general public how and when various medical discoveries were made and how that information affected health care of the time period. The set starts with primitive humans and concludes with a final volume that presents readers with the very vital information they will need as they must answer society's questions of the future about everything from understanding one's personal risk of certain diseases to the ethics of organ transplants and the increasingly complex questions about preservation of life.

Each volume is interdisciplinary, blending discussions of the history, biology, chemistry, medicine and economic issues and public policy that are associated with each topic. *Early Civilizations,* the first volume, presents new research about very old cultures because modern technology has yielded new information on the study of ancient civilizations. The healing practices of primitive humans and of the ancient civilizations in India and China are outlined, and this volume describes the many contributions of the Greeks and Romans, including Hippocrates' patient-centric approach to illness and how the Romans improved public health.

The Middle Ages addresses the religious influence on the practice of medicine and the eventual growth of universities that provided a medical education. During the Middle Ages, sanitation became a major issue, and necessity eventually drove improvements to public health. Women also made contributions to the medical field during this time. *The Middle Ages* describes the manner in which medieval society coped with the Black Death (bubonic plague) and leprosy, as illustrative of the medical thinking of this era. The volume concludes with information on the golden age of Islamic medicine, during which considerable medical progress was made.

The Scientific Revolution and Medicine describes how disease flourished because of an increase in population, and the book describes the numerous discoveries that were an important aspect of this time. The volume explains the progress made by Andreas Vesalius (1514–64) who transformed Western concepts of the structure of the human body; William Harvey (1578–1657), who studied and wrote about the circulation of the human blood; and Ambroise Paré (1510–90), who was a leader in surgery. Syphilis was a major scourge of this time, and the way that society coped with what seemed to be a new illness is explained. Not all beliefs of this time were progressive, and the occult sciences of astrology and alchemy were an important influence in medicine, despite scientific advances.

Old World and New describes what was happening in the colonies as America was being settled and examines the illnesses that beset them and the way in which they were treated. However, before leaving the Old World, there are several important figures who will be introduced: Thomas Sydenham (1624–89) who was known as the English Hippocrates, Herman Boerhaave (1668–1738) who revitalized the teaching of clinical medicine, and Johann Peter Frank (1745–1821) who was an early proponent of the public health movement.

Medicine Becomes a Science begins during the era in which scientists discovered that bacteria was the cause of illness. Until 150 years ago, scientists had no idea why people became ill. This volume describes the evolution of "germ theory" and describes advances that followed quickly after bacteria was identified, including vaccinations, antibiotics, and an understanding of the importance of cleanliness. Evidence-based medicine is introduced as are medical discoveries from the battlefield.

Medicine Today examines the current state of medicine and reflects how DNA, genetic testing, nanotechnology, and stem cell research all hold the promise of enormous developments within the course of the next few years. It provides a framework for teachers and students to understand better the news stories that are sure to be written on these various topics: What are stem cells, and why is investigating them so important to scientists? And what is nanotechnology? Should genetic testing be permitted? Each of the issues discussed are placed in context of the ethical issues surrounding it.

Each volume within the History of Medicine set includes an index, a chronology of notable events, a glossary of significant terms and concepts, a helpful list of Internet resources, and an array of historical and current print sources for further research. Photographs, tables, and line art accompany the text.

I am a science and medical writer with the good fortune to be assigned this set. For a number of years I have written books in collaboration with physicians who wanted to share their medical knowledge with laypeople, and this has provided an excellent background in understanding the science and medicine of good health. In addition, I am a frequent guest at middle and high schools and at public libraries addressing audiences on the history of U.S. presidential election days, and this regular experience with students keeps me fresh when it comes to understanding how best to convey information to these audiences.

What is happening in the world of medicine and health technology today may affect the career choices of many, and it will affect the health care of all, so the topics are of vital importance. In addition, the public health policies under consideration (what medicines to develop, whether to permit stem cell research, what health records to put online, and how and when to use what types of technology, etc.) will have a big impact on all people in the future. These subjects are in the news daily, and students who can turn to authoritative science volumes on the topic will be better prepared to understand the story behind the news.

ACKNOWLEDGMENTS

This book and the others in the series were made possible because of the guidance, inspiration, and advice offered by many generous individuals who have helped me better understand science and medicine and their histories. I would like to express my heartfelt appreciation to Frank Darmstadt, whose vision and enthusiastic encouragement, patience, and support helped shape the series and saw it through to completion. Thank you, too, to the Facts On File staff members who worked on this set.

The line art and the photographs for the entire set were provided by two very helpful professionals. The artist Bobbi McCutcheon provided all the line art; she frequently reached out to me from her office in Juneau, Alaska, to offer very welcome advice and support as we worked through the complexities of the renderings. A very warm thank you to Elizabeth Oakes for finding a wealth of wonderful photographs that helped bring the information to life. Carol Sailors got me off to a great start, and Carole Johnson kept me sane by providing able help on the back matter of all the books. Agent Bob Diforio has remained steadfast in his shepherding of the work.

I also want to acknowledge the wonderful archive collections that have provided information for the book. Without places such as the Sophia Smith Collection at the Smith College Library, firsthand accounts of the Civil War battlefield treatment or reports such as Lillian Gilbreth's on helping the disabled after World War I would be lost to history.

INTRODUCTION

Perhaps the history of the errors of mankind, all things considered, is more valuable and interesting than that of their discoveries.

—*Benjamin Franklin*

At the beginning of the 18th century, the world was poised for great change. Societal shifts—from changes in attitude and a willingness to question to the very real expansion of geographic boundaries—were occurring everywhere. The revival of interest in ancient Greece and Rome that began in the 1500s eventually broke the dominance that the Catholic religion had exerted in western Europe during the Middle Ages, and this enabled people to ask new questions and be open to alternative answers. Explorers were returning from lands never imagined with reports of peoples and samples of plants and other products that sparked scientists, physicians, and merchants to look for new uses for both old and new substances. Agricultural techniques improved, requiring fewer people to grow food and freeing more to migrate to cities.

Science was forever changed by the realization that Aristotle's method of *deductive reasoning* (a process that accepts a hypothesis and then builds the case to support it) was less helpful with "real life." Galileo and his contemporaries realized, in nature and in human health, it was enormously difficult to determine "simple true statements" about how things worked. With a major push from Sir Francis Bacon, the scientists of the day adopted *inductive reasoning*, which is the deductive method in reverse. In inductive reasoning, a scientist starts with many observations of nature and through this fact-gathering process creates observations that can be tested to prove how nature works. Since 1600, the inductive method has been incredibly successful in investigating nature, surely far more successful than its originators could have imagined. The inductive method of investigation has become so entrenched in science that it is often referred to as the scientific method.

In addition, scientific progress in specific fields was leading to new developments. From practical "aids" such as the development of the steam engine (1698) to inevitable advances like new—and more treacherous—ways to use gunpowder, change was occurring rapidly. And as the relatively new invention of the printing press began to be used for printing information beyond books, such as broadsides and maps and pamphlets with medical advice, an increasing amount of material was being written in the vernacular, making the information available to more and more people. This era was known as the Enlightenment, a time when skepticism ruled and experimenting was smiled upon and philosophers believed that reason would trump all.

This period of history also brought about impressive achievements in scientific understanding that pertained to medicine. The documentation of anatomy progressed rapidly, and there was greater understanding of how certain bodily systems worked (the workings of the vascular system being the most notable). Scientists expressed great hope for the future, with some proclaiming that because scientific knowledge was growing so rapidly disease would soon be completely eradicated. Yet the actual cause of illnesses still stymied them. Though microscopes provided the capability of seeing "little animalcules," no one had drawn a line between the presence of what we now know as bacteria and disease. Also, there were many misguided theories about cause and effect, which led scientists down paths that kept them from more relevant discoveries.

Old World and New: Early Medical Care, 1700–1840 illuminates what occurred during the Enlightenment to affect future developments in medicine. The back matter of the volume contains a chronology, a glossary, and an array of historical and current sources for further research. These sections should prove especially helpful for readers who need additional information on specific terms, topics, and developments in medical science.

Hindrances to good health came about with the increasingly mobile society. As worldwide trading grew, diseases began to spread widely and often virulently, with waves of epidemics greatly worsening mortality rates of the period. Warfare erupted frequently, leading to an increase in death and a greater number of wounded, and the Industrial Revolution led to a greater number of injuries to workers as well as higher levels of air pollution and waste runoff that affected the area's water. As population in urban areas grew, the health problems posed by poor sanitation increased, and the urban poor suffered particularly. Despite the fact that the scientific method was beginning to lead to new learning, it would take a long time before it had an impact on people's health.

Chapter 1 provides a fascinating look at the approaches to medicine that were used early in the 1700s. Mesmerism, *phrenology,* and bloodletting were the fads of the day, and these methods will be explained. The advances made in midwifery, anatomy, and surgery are outlined in chapter 2; the Chamberlen "family secret" (forceps) made a big difference in the birth experience, and its story is particularly fascinating. Chapter 3 outlines the major progress made in battlefield medicine. For the first time, soldiers were treated on the battlefield or carried off the battlefield by medical personnel, and this was to make a big difference in survival rates. In an era when physicians had no idea how disease spread, they still made valiant efforts to keep people from catching diseases, and that topic will be discussed in chapter 4. Chapter 5 is a focused look at yellow fever, its impact on the nation's then capital (Philadelphia), and how physicians went about treating it. Chapter 6 outlines early American medical care, from the qualifications of the physicians who practiced medicine to what became a major industry—*patent medicines.* No one really understood the digestive process until this era, and how they learned about the way the body processes food is an interesting story that is presented in chapter 7. Chapter 8 addresses the very important issue of public health. Communities were beginning to realize how important it was that they take organized steps to improve the health of their citizens, and this chapter will explain what they did.

This book is a vital addition to the literature on the history of medicine because it puts into perspective the medical discoveries of the period and provides readers with a better understanding of the accomplishments of the time. While physicians of this era did not yet know the cause of disease, they had begun making many advances that were to be key to medical improvements to come.

1

Medicine in Search of Better Answers

The advent of the 18th century brought about what is referred to as the Enlightenment, a period when the educated and professionals among the citizenry of Europe were willing to question old standards in everything from religion to science. In the process, this elite group began turning away from much of the beliefs in the magical or miraculous and toward a more scientific and rational way of thinking. Lifestyles were changing, and this, too, spurred new progress. As farming methods improved, fewer people were needed to produce food, so more people migrated to the cities where industry was expanding and offering an increasing number of jobs. Those who did not work producing goods were able to find employment as shop attendants or in other service positions that were needed because of the growing urban population. In addition, explorers were actively traveling the globe, and colonial settlements were popping up in distant parts of the world. Trade with people on faraway continents introduced new plants for food and medicines, while at the same time importing new diseases.

In the world of medicine, the discovery about the workings of the heart by William Harvey and the advancement in anatomical studies by Vesalius in the preceding era had created a new baseline for medicine. Optimism about this new progress prevailed, and there were predictions from scientists that all disease would one day be eradicated. However, this was overly optimistic, as they were still a very long way from any sort of understanding about what caused disease and continued allegiance to the theory that correcting imbalances would cure almost everything.

This chapter highlights several of the most popular healing methods of the time. Bloodletting remained a frequently used process for restoring health, and two pseudoscientific doctrines relating to medicine emerged from Vienna in the latter part of the century and attained great popularity. Mesmerism was a mode of healing introduced by Franz Anton Mesmer that involved magnetization (and the power of Mesmer's presence). Phrenology was another system that was explored as a serious effort to better understand the workings of the brain. These approaches did not prove to be medically significant, but their invention, their acceptance, and their ongoing use provide deeper understanding of the state of medicine in the 18th century.

During this era, patients were more likely to be overtreated than undertreated. In this time of excessive bloodletting, Samuel Hahnemann developed the concept of *homeopathy,* and his method will be explained.

DIFFERENT VIEWS ON RESTORING HEALTH

Almost all scientists and physicians who lived during this time believed that the root of all problems lay in imbalances. There were many lines of thought as to what was out of balance and what do to about it. Physicians and scientists from all over Europe came up with differing theories as to how to bring about cures.

The Dutch professor and physician Herman Boerhaave (1668–1738) continued to be very influential; his 1708 textbook *Institutiones medicae* was an important contribution to medicine. Boerhaave will long be appreciated for bringing up to date and organizing the medical information of the time and for realizing that medical students would learn a great deal more if some of the lessons took place at patients' bedsides. However, his overall explanation of the causes of disease was misguided. This well-respected professor believed bad health resulted from mechanical imbalances, and he classified disorders in two categories: Some resulted from an imbalance of "solids" (like tuberculosis); others resulted from an imbalance of "blood and humours." (Blood clots were an example of this type of imbalance.) Cures were usually both simplistic and ineffective. Bloodletting was used to rebalance "solid" imbalances; milk and iron were prescribed for other types of problems.

German physician and chemist George Ernst Stahl (1660–1734) believed that "animism" (the soul) was at the heart of everything. He felt that the psyche directed the body and regulated physiology. For Stahl, disease was the soul's attempt to reestablish bodily order. Others had different theories. Boissier de Sauvages (1706–67) believed the body was a machine, and disease was nature's effort to expel "morbid" matter. In Edinburgh, John Brown (ca. 1735–88), a Scottish physician and essayist, believed that all diseases could be classified as either increasing or decreasing "excitement" on the body, and that treatments should be planned accordingly. He was among the few who spoke out against bloodletting, and his chief remedies were alcohol and opium.

Scottish professor of medicine William Cullen (1710–90) and his followers believed that life was a function of nervous energy, and disease was disturbance in this life force, but he felt that disease could be classified in the following four major divisions:

- febrile diseases
- neuroses, or nervous diseases
- diseases produced by bad bodily habits
- local diseases

Cullen, however, fully acknowledged that an understanding of how disease transferred to a person was still missing: "We know nothing of the nature of contagion that can lead us to any measures for removing or correcting it. We know only its effects."

Though vigorous debate continued about the cause and classification of disease, one opinion was shared, and that was for the need for treatment.

PREFERRED METHOD OF TREATMENT: BLOODLETTING

Because imbalances were still thought to be a large part of any disease, bloodletting remained firmly entrenched. Medicinal bloodletting had been practiced since the Stone Age when early humans believed that evil spirits could be removed through the drawing of blood. The practice became even more prevalent when the ever-influential Galen, the second-century Greek physician, theorized that having a *plethora* of blood was dangerous and caused bad health. By the 1700s, bloodletting was commonly used as part of regular health maintenance as well as a healing method to remove "useless" blood to remedy inflammation, hemorrhaging, fevers, and a multitude of other illnesses.

As physicians continued to recommend bloodletting, the "art and science" of it became increasingly important, and physicians formulated specific methods they felt would make the bloodletting more effective. Physicians decided that the proper site for the blood draw depended on the problem that was to be cured and made very serious evaluations as to the quantity of blood to be removed. As the medical professionals began to understand more about the circulatory system, arguments about the selection of the proper site increased. Some physicians believed that the site of the bloodletting should be on the side opposite the lesion. Others chose a site close to the source of the problem "in order to remove putrid blood and attract good blood" to the area. The amount to be removed was carefully decided and just as carefully measured as a vein was opened and blood flowed into a bowl.

The use of *leeches* was popular during this period, and this practice, too, had ancient origins. Egyptian tombs show illustrations of the use of leeches from as early as 1500 B.C.E., and the Chinese write about it in documents dating to the first century C.E. Galen often recommended the use of leeches to correct any imbalance of the four humors, and physicians used leeches for several thousand years because they believed it could remedy complaints like headaches and *gout.* Just as the physician gave careful thought to the opening of a vein, they also carefully considered how and where to use leeches. They felt they had some understanding of how much each leech consumed, so they prescribed both the number of leeches to be used and where they should be placed. Leeches drop off when they are full, so this gave surgeons an added feeling of confidence in using them.

Leeches were used for epilepsy, hemorrhoids, obesity, tuberculosis, and headaches. If a person was suffering a particularly debilitating headache, leeches were applied inside the nostrils. Edinburgh surgeon John Brown (1810–82) wrote of treating himself for a sore throat by having six leeches and a mustard plaster placed on his neck. Then he had a dozen leeches placed behind his ears, and he reported removing 16 ounces (0.47 l) of blood by *venesection.*

Leeches were so popular that apothecary shops kept a bowl filled with live leeches, just as they kept on hand other medicinal mixtures. By the early part of the 19th century, the use of leeches reached a peak. The type of leech that was so popular in Europe, the *Hirudo medicinalis,* had been hunted to extinction so they had to be imported. In North America, a native type of leech was available, but American diseases were thought to be particularly virulent, and the European leech was considered to be more effective so there was also an American market for the *Hirudo medicinalis.*

By the 1830s, the practice of bloodletting finally declined somewhat. Physicians were beginning to see that patients who were bled did not necessarily recover more fully than those who underwent other treatments.

Modern Day Medicinal Use of the Leech

In 1884, John Haycraft, a professor of physiology at the University of Wales, made an interesting discovery. He found that blood in the leech gut did not *coagulate.* While this information had no practical use at the time, scientists frequently benefit much later from the small discoveries made by those who preceded them, and this was what occurred here. In the 1950s, a German scientist Fritz Marquardt was able to isolate the anticoagulant hirudin from leech pharyngeal glands, and, with this additional knowledge, physicians began to reintroduce the possibility of using leeches in medicine.

By the 1980s, there were two groups of surgeons who found that leeches were particularly helpful in their work. The first group were doing complex reattachment surgery—reattaching fingers and ears or large sections of scalp, all of which depend on reestablishing blood flow to an area—and this is not easy to accomplish. The surgeon needs to reconnect as many arteries and veins as possible, and, if not enough veins are reconnected, the blood may pool in the reattached organ or limb, preventing fresh, oxygenated blood from entering and restoring health to the reattached part.

The second group who were excited about the possible promise of leeches were plastic surgeons who face a similar dilemma, needing to restore blood flow to an area. Physicians learned that when leeches are applied to tissue they remove blood and secrete several compounds that have *vasodilator,* anticoagulant, and clot-dissolving properties. This prevents the tissue from dying and allows the body to reestablish healthy blood flow to the reattached part. Leeches are also particularly helpful for use with burn victims in attaching donor skin flaps to new areas as they drain blood clots and improve adhesion to a recipient site. The use of leeches has known risk, however. About 20 percent of surgical patients who receive leech treatment get infections. These infections can generally be prevented if antibiotics are given.

In a continuing effort to find other ways that leeches might be helpful, doctors have been experimenting with using them to reduce knee pain in people with arthritis. Additional studies are needed before this becomes an accepted practice.

In 2004, the Food and Drug Administration granted a French firm permission to market leeches for medicinal purposes. In the future, scientists anticipate the use of live leeches will probably be replaced by a synthetic drug. The new techniques of molecular biology coupled with pharmaceutical companies' interests in benefiting from this type of invention probably spells the end of the use of live leeches.

MESMERISM BECOMES A POPULAR METHOD OF HEALING

Physician Franz Anton Mesmer (1734–1815) popularized a healing method that was based on his belief that all things were connected by a subtle and mobile fluid that pervades the universe. When this fluid within the body is blocked, the result is disease. Mesmer became quite well known, and, although leading scientists investigated and rejected his theory, it had no effect on his popularity.

Mesmer began his medical studies in 1759 at the University of Vienna. His doctoral dissertation, which borrowed heavily on the work of an English physician and friend of Newton's, concerned the influence of the planets on the human body and disease, a subject that was to be the underpinning of much of Mesmer's work. Mesmer believed that people contained *magnétisme animal,* and it was this force that had to be kept flowing smoothly. (This term is frequently translated as "animal magnetism," which today refers to sex appeal. Experts stress that this is an incorrect translation. The term *animal,* as Mesmer used it, refers to the Latin word *animus,* meaning "breath" or "life force.")

Mesmer's early work was influenced by a Jesuit healer who "cured" people using a magnetic plate. Mesmer theorized that this worked because of the universal fluid that flowed through everything. As early as 1774, Mesmer had a young patient swallow a preparation containing iron, and then Mesmer attached magnets to various parts of her body. The patient reported that she could feel a mysterious fluid running through her body, and for several hours she was relieved of her symptoms. Mesmer claimed he had restored the flow of *magnétisme animal* and in so doing had brought about a cure.

Contemporary painting of 18th-century mesmerism *(Massimo Polidoro)*

This treatment method had its complications, so Mesmer worked to simplify the process. When healing one-on-one, he sat directly in front of the patient with his knees against the patient's knees. Pressing one of the patient's thumbs into the palm of his own hand, Mesmer gazed into the patient's eyes and used his other hand to wave over the patients' shoulders. Sometimes he waved a magnetized pole over the patient, but increasingly he used only the power of his hands to help restore the movement of the bodily fluid. Sometimes, he also pressed his fingers into the area below the patient's diaphragm for as long as an hour or two. Treatments generally concluded with music, but it should not be assumed that these sessions ended peacefully. Mesmer believed that when ill, a person got worse before getting better, so a fever might have gone higher or an insane person might have experienced a fit of madness before showing signs of improvement.

Mesmer's first practice was in Vienna, but he soon was forced to move on. He claimed he had restored the sight to a blind musician, but when this claim proved false, the community turned against him. Mesmer moved to Paris in 1778 and established a practice in a wealthy neighborhood. He hoped for the approval of a scientific organization to add credibility to his work, but his requests to the Royal Academy of Sciences and the Royal Society of Medicine were both turned down. However, a well-respected French physician Charles d'Eslon became his follower, which added to Mesmer's credibility.

Mesmer soon had a waiting list of patients so he devised ways to treat groups of people collectively. One of the methods enabled him to heal as many as 20 people at a time. For this purpose, Mesmer created a large barrel-like vessel (Mesmer called it a *baquet*) that had iron rods of different heights attached. The *baquet* was filled with glass bottles arranged in a radial pattern sitting in water on a bed of crushed glass, pounded sulfur, and iron filings; each rod had magnets at its base. The patients were each assigned to hold onto one of the rods, and the rod was selected for the patient based on area of illness—those with headaches were given a rod that could be placed against the head; those with gastric distress used a rod that could be placed on their midsection, etc. (When Mesmer updated this invention he added a Leyden jar that served

(continues on page 12)

THE SALEM WITCH TRIALS

A fear of witchcraft pervaded both Europe and the colonies, and communities often accused healers of practicing sorcery. A local healer—almost always a woman—was victimized, perhaps because of an unfortunate result with one of her patients or sometimes because a community simply turned against her. The healer would be suspected, accused, put through a trial of sorts, and beheaded, hung, burned, or drowned, depending on the spirit of the times.

The Salem witch trials, known for many reasons including *The Crucible,* Arthur Miller's award-winning 1954 play, are significant to the history of medicine because they exemplify the attitude specific to that era, toward illness and healing. In Salem, the accused were not actually the town's healers. They were young children who were exhibiting odd behavior, and the town healer determined that they were possessed.

Salem witch trials

To fully understand the circumstances, it is helpful to know that during the late 17th century, life in Salem was very unsettled. At that time, the Massachusetts colony (of which Salem was a part) lacked both a charter and a governor, so there was a high level of general uneasiness. Some members of the Puritan community were also very upset about the group's selection of their first ordained minister. When the minister's daughter, the minister's niece, and another young girl, all between the ages of nine and 12, began exhibiting strange behavior (screaming, throwing things, complaining they were being cut with knives, etc.), it was very unnerving to the people of the community. At that time, the possibility of witchcraft was plausible. The Puritans believed that God punished all sinners with illness or calamity, and physicians were sometimes pressured into these types of diagnoses, particularly if the beliefs of the patient's family and friends were so strong that they were convinced that the person was possessed.

In most communities, the first call for help at this time would have been to a woman healer. Salem, however, had a man who served as the town physician though his exact training is unknown. When the physician Dr. William Griggs could find no physical ailment to explain the girls' behavior, he determined that they were "under an Evil hand." Dr. Griggs expressed powerlessness to help the girls, so the community looked outward for the witches who must have been causing the spell. When the girls' conditions seemed to worsen, the community focused on an Irish washerwoman and a female slave as the likely suspects. However, the community outrage did not stop there. Over the next few months, 20 people (14 women and six men) were executed for witchcraft, and between 175 and 200 people were imprisoned, including a

(continues)

(continued)

four-year-old girl who, without awareness, implicated her mother. The little girl then confessed to being a witch herself so that she would not be separated from her parent.

Recent scholars have offered various theories as to the girls' behavior. Many felt it was hysteria, though a few have looked for illness that might have caused their reactions. One scientist posited that the girls might have ingested bread that was made of moldy grain; one particular kind of mold contains a chemical compound that can have a hallucinogenic effect. Most scientists discount this theory, but one book offers another idea. *A Fever in Salem* by Laurie Winn Carlson suggests that those who acted bewitched may have suffered from encephalitis lethargica, a disease with somewhat similar symptoms. Still others have suggested that the girls had Huntington's chorea.

The theories will undoubtedly continue to be explored, but in the meantime there is no diminishment of fascination with the witches of Salem. The community reactions were not surprising considering the lack of information about diseases, about healing, or about any kind of mental problems—including hysteria—from which the girls may have been suffering.

(continued from page 9)
as a battery, and the force emanated through the metal rods. When a patient touched the rod, they got a jolt, which must have been presented as part of the cure.)

While Mesmer surely understood that he was running a profitable business, he took into account those who could not pay for his services. At the Hôtel Bullion, he had three rooms set aside for treating paying guests and one additional room for people who could not afford to pay him. Patients were assigned to their rooms and given specific places to sit. After everyone was in place, Mesmer entered the room wearing a flowing lilac-colored silk robe and ornate gold slippers. His arrival often sent people into hysterics. If someone became very upset, Mesmer's valet had to remove the person to what was a soundproof, padded room; frequently several in the group had to be removed.

His business grew to the point that he was seeing 200 people or more a day, yet the demand still exceeded what he could satisfy. The press referred to it as "Mesmeromania." To satisfy patients—and line his pockets—Mesmer created and sold a *baquet* for home use, and he started the Society of Universal Harmony where he trained others in his method. By 1789 (the beginning of the French Revolution), the society had graduated 480 people. Mesmer also magnetized several trees around Paris and announced that they, too, were healing.

Mesmer appealed to the high society of the day, and the king's wife attended sessions. When rumors began that hinted that Mesmer was actually conducting sexual orgies, the king was compelled to investigate. Louis XVI decided that the best person to investigate was Benjamin Franklin, a well-respected American who understood science and was currently in Paris as the American ambassador to France. Antoine-Laurent de Lavoisier (see chapter 7), Jean-Sylvain Bailly, and Dr. Joseph-Ignace Guillotin, who created the guillotine (the device that would later bring an end to both Lavoisier and Bailly), were asked to serve on the panel with Franklin. Because Mesmer was so popular, the king did not charge the panel with investigating Mesmer's practice. They were simply charged with examining whether a new force in physics had been discovered, as Mesmer claimed.

Mesmer was alarmed by the investigation and refused to cooperate, so the panel worked with his devoted follower Dr. D'Eslon. The first method the men devised for testing mesmerism was to experience the process. Franklin had long suffered from gout, so he attended a session but reported no improvement. Next they examined the healing process of magnetized trees. One tree in Franklin's garden was magnetized, and a young boy, age 12, was asked to identify the one that had the strongest magnetic force. The boy walked toward a tree, claimed he felt the force, and promptly fainted. Unfortunately for Mesmer, the boy had gravitated toward the wrong tree.

Though discredited, mesmerism continued to be used in various forms and with different practitioners. In 1841, Dr. James Braid (1795–1860) witnessed mesmerism and developed what is now known as hypnotism. (The system used by Mesmer did not involve hypnotism although it is sometimes indicated that it did.)

SCIENCE ADVANCES THROUGH PHRENOLOGY

In the 18th century, the function and the workings of the brain were a mystery. When Franz Joseph Gall (1758–1828) began studying the skull and how it related to the brain—what became known

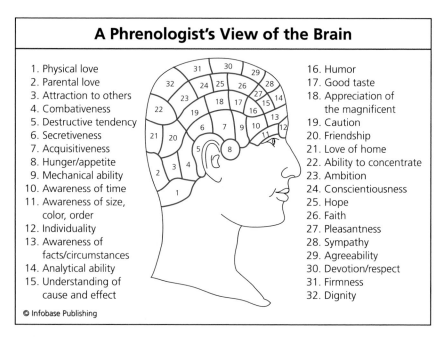

A Phrenologist's View of the Brain

1. Physical love
2. Parental love
3. Attraction to others
4. Combativeness
5. Destructive tendency
6. Secretiveness
7. Acquisitiveness
8. Hunger/appetite
9. Mechanical ability
10. Awareness of time
11. Awareness of size, color, order
12. Individuality
13. Awareness of facts/circumstances
14. Analytical ability
15. Understanding of cause and effect

16. Humor
17. Good taste
18. Appreciation of the magnificent
19. Caution
20. Friendship
21. Love of home
22. Ability to concentrate
23. Ambition
24. Conscientiousness
25. Hope
26. Faith
27. Pleasantness
28. Sympathy
29. Agreeability
30. Devotion/respect
31. Firmness
32. Dignity

© Infobase Publishing

Based on the work of Franz Gall, phrenology became very popular. It was thought that a person's skull could be felt carefully by a trained phrenologist who could then explain that person's personality traits.

as phrenology—it caught the interest of the scientific community. Over time, he theorized that the contours of the skull were a guide to an individual's mental faculties and character traits, and this idea became quite popular. While Gall's theory was misguided, his work is notable because he was one of the first to consider the brain to be the source of all mental activity.

The practice of phrenology came to involve feeling the bumps in the skull to determine a person's psychological attributes. Phrenologists were trained to run their fingertips and palms over the skulls of their patients to feel for enlargements or indentations. Gall taught that each bump or indentation was indicative of a specific trait ranging from the ability to see well to the likelihood of being very devout.

Gall's Life

Franz Gall was born in Baden, Germany, and as a second-born son he traditionally would have joined the priesthood. Gall decided to study medicine instead, and he began to develop his theory that each person's external characteristics were symptomatic of individual talents. He concluded that prominent eyes indicated a powerful memory, and Gall came to believe that the brain consisted of 27 different "organs," each corresponding to a discrete human faculty. The cranial bone conformed in order to accommodate the different sizes of these particular areas of the brain in different individuals, so a person's capacity for a particular personality trait could be determined by using a *caliper* to measure the part of the skull that covered that part of the brain. Gall eventually created a mind map with traits and corresponding numbers. As a phrenologist felt the skull, he could refer to a numbered diagram showing where each functional area was believed to be located.

Gall's ideas were condemned by both science and religion. Science felt he had no proof; religion felt the theory was contrary to a belief in God. Because of this disapproval, Gall left his teaching position in Austria in 1805 with colleague Johann Spurzheim (1776–1832). He scheduled lectures elsewhere in Europe that were very well attended, and in 1807 Gall and Spurzheim decided to settle in Paris. He unsuccessfully sought admission to the Academy of Sciences. Then, Gall was denounced by Napoléon, the ruling emperor, and this made Gall a celebrity among the intellectual class, who embraced him.

Though Gall started a book about the subject and the first two chapters were made public in 1791, he never completed it. However, he did many lectures on the topic and also wrote a detailed letter about his ideas. Johann Spurzheim took Gall's theory of cranioscopy and refined it into a system that he taught regularly and came to call phrenology. Spurzheim successfully spread the ideas of phrenology to the United Kingdom and the United States where the term was popularized.

While phrenology was rejected by mainstream academia, many people still consulted phrenologists. It varied from being a fairground entertainment to being written about as serious science. The phrenological analysis was used to predict the kinds of relationships and behaviors to which a patient was prone. In its heyday from the 1820s to the 1840s, phrenology was often used to predict a child's future, to assess prospective marriage partners, and to provide background checks for job applicants.

Spurzheim was not Gall's only follower; he influenced many. Two Scottish brothers George Combe (1788–1858) and Andrew Combe (1797–1847) founded the Phrenology Society of Edinburgh, and George wrote a lot on phrenology. In addition, two American brothers Lorenzo Niles Fowler (1811–96) and Orson Squire Fowler (1809–87) started a phrenological firm and publishing house devoted to information on phrenology. In the United States, the Reverend Henry Ward Beecher actively promoted it as a source of psychological growth, and it caught on with the British prime Minister David Lloyd George as well.

Though the theory makes little sense today, Gall made significant contributions to neurological science. He was the first to suggest that character, thoughts, and emotions were not located in the heart but in the brain. Because phrenology was so popular, this became a belief that was much more widely accepted than it

THE PSYCOGRAPH

Phrenology was largely abandoned in Europe by the latter part of the 19th century, but there were still believers elsewhere who devoted time and energy to this "science." American businessman Henry C. Lavery and a partner Frank P. White from Superior, Wisconsin, took their life savings—White contributed $39,000 that had been invested in the 3M Company—to finance research to create a machine that would evaluate a person's skull to assess personal characteristics.

Twenty-six years later they had a machine they were satisfied with. The Psycograph consisted of 1,954 parts, including a hoodlike device with 32 probes that touched the person's head. The machine was preloaded with 160 possible statements, and the "score" the person received varied since there were an almost unlimited number of possible combinations. To be analyzed, the subject sat in a chair connected to a headpiece (like hair dryers at beauty salons). When the operator pulled the lever to activate the machine, the probes sent back signals that identified appropriate statements for each part of the head and its related trait. The operator then presented the person with the "personal diagnosis."

According to Bob McCoy, curator of the Museum of Questionable Medical Devices, Lavery and White's company built 33 machines that were leased to entrepreneurs throughout the country for a $2,000 down payment plus $35 a month rent on the machine. They were popular attractions for theater lobbies and department stores, as they were good traffic builders during the Great Depression, and business flourished. In 1934, two entrepreneurs set up a machine at the Chicago Exposition, the Century of Progress, and they netted $200,000!

otherwise would have been, and it paved the way for new thinking on this topic.

HOMEOPATHY COMES INTO VOGUE

In a day when "no treatment" or "too little treatment" was tantamount to "neglect," physicians operated with the understanding that the more they did the better off the patient would be.

Healing in the 18th century often consisted of multiple bloodlettings alternating with purging of the body, and pharmaceutical "prescriptions" of the time still involved medicines like Galen's Venice treacle, that was made of 64 substances including opium, myrrh, and viper's flesh. People often died from the treatment process alone. George Washington was among those who probably died of his cure. (See chapter 6.) With the supervision of three distinguished physicians, Washington was bled, purged, and blistered until he died about 48 hours after complaining of a sore throat (1799).

At a time when prescriptions were lengthy and doses were large, Samuel Hahnemann (1755–1843), a German physician, presented a system that was in stark contrast to the "more is better" philosophy, and his system of homeopathy gained many followers. Homeopathy (*homeo* for "similar," *pathy* for "disease") is a natural, noninvasive system of medical treatment involving minute doses of drugs, and it was based on the theory that substances that cause certain symptoms in a healthy person can—in small amounts—cure those symptoms in a person with the disease. Homeopathy is often incorrectly used to refer to any form of alternative medicine, and while true homeopathy takes a holistic approach to health, it is designed to help the body heal itself, not to suppress or control symptoms.

Samuel Hahnemann was born in Saxony to a family of porcelain painters, but he showed a remarkable ability in languages, and, by age 20, he was said to have mastered English, French, Italian, Greek, and Latin, and he translated and taught while gaining his medical education. After settling with his wife in a mining area of Saxony, Hahnemann gave up practicing medicine in 1784 because he was concerned that the treatment methods were cruel and ineffective. His ability with language and his knowledge of medicine made him the perfect candidate to translate the famed Edinburgh professor William Cullen's lectures on the *Materia medica* into German (see chapter 2 for more about William Cullen). Hahnemann was puzzled by Cullen's comment that cinchona was effective at treating malaria because it was an astringent, but Hahnemann, who often wrote about chemistry, knew that malaria did not respond to other forms of astringents. This led him to experiment on himself by taking some of the cinchona bark to see what happened. He observed that he soon exhibited symptoms similar to the disease. Hahnemann experimented with other substances and determined that "like cures like" (*similia similibus curantur*). This led him to develop a homeopathic principle that a disease could be cured by a drug that would produce in a healthy person symptoms similar to that of the disease (1796). Hahnemann concluded that by using a drug to create artificial symptoms, this empowered the vital force of the body to neutralize and expel the original disease.

In working out his theory, Hahnemann deduced that the key to stimulating healing rather than a toxic response had to do with using minute amounts of the drugs diluted in other compounds. In 1810, Hahnemann explained his theory in *Organon of the Medical Art*. He returned to Leipzig to teach his new medical system but was soon driven out of town by irate apothecaries who disliked his advocacy of a reduction in the use of drugs. He re-settled in Paris where he continued to practice homeopathy undisturbed.

Homeopathy did not become popular in America until 1825, and its rise in the United States will be discussed in chapter 6.

CONCLUSION

While this era began with many elements that involved clinging to the clinical methods of the past—mysticism and the fear of witchcraft among them—there were also early signs of sound scientific thinking in physics, chemistry, and the biological sciences. These were converging to form a rational scientific basis for every branch of clinical medicine. Men like Franz Gall and Samuel Hahnemann may not have had the correct answers to medical assessment and healing, but they were beginning to lead the way so that new doors would open and progress in clinical medicine could continue.

2

Advancements in Midwifery, Anatomy, and Surgery

By the 18th century, medical schools were becoming much more common. The University of Leiden in the Netherlands had grown to rival the one in Padua, and more universities were adding a medical curriculum to their regular offerings. After Scot John Monro attended Leiden where he prepared to become a surgeon in the army, he returned to Edinburgh determined that Edinburgh should soon have a medical program. John still sent his son Alexander to Leiden (where he studied under the well-respected Hermann Boerhaave), but when Alexander returned, he was appointed professor of anatomy at Edinburgh. This was the start of what became a very well-regarded medical program responsible for educating many of the leaders in the field.

During the 17th century, female midwives continued to be the mainstay of childbirth, even though male physicians were sometimes consulted when the circumstances were particularly difficult. This custom began to change as men opted to come into the field. Two of the most prominent of this period were William Smellie and William Hunter, whose brother John Hunter went on to excel in the field of surgery. This was an area that was beginning to expand, although there was still no anesthesia to numb pain and no one understood the importance of sterility in guarding against infection.

Anatomy was not seen as vital to the teaching of medicine and thus, students who wanted to study the body had to sign up for classes from a private institution. Even then there were challenges—cadavers were difficult to come by. During the 17th century, anatomy was studied by a very small group, so the need for bodies could be more or less met by judges willing to sentence some people to "death and dissection." (One judge who admired the work of Vesalius, the great anatomist, timed his executions to suit the convenience of Vesalius's dissection schedule. More executions were scheduled for the winter months because the cold weather helped preserve the remains for longer.)

During the 18th century, two unrelated trends began that affected the availability of bodies for study. Just as the number of medical and private anatomy schools began to increase, Europe experienced a drop in capital punishment. By the 19th century, statistics showed that in the United Kingdom only 55 people were executed one year, and yet, with the expansion of medical schools, as many as 500 bodies were needed each year. As explained later in the chapter, body snatching became a necessary part of the process.

This chapter will discuss the changing practices in midwifery and the art of surgery as practiced by the Hunter brothers, as well as the methods used for procuring bodies for anatomical study.

MIDWIFERY BEGINS TO CHANGE

While women had long turned to other women for help during childbirth, this began to change during the mid to late 1600s. There were still serious issues to consider—men were still not supposed to see women disrobed even for medical reasons and they did not really have a lot of helpful information since women had for so long been the ones who handled childbirth. However, the practice of a male physician standing by to advise the midwife during a difficult childbirth must have established them as authority figures who could be helpful.

One of the women who felt strongly about having a physician present was the duchess de La Vallière, who was at the court of Louis XIV when she gave birth in 1663, according to material provided by Alastair McIntyre, director of the Scottish Studies Foundation. Because she was fearful of what was about to happen to her, she called in Julien Clément, a well-respected surgeon. Clément was secretly ushered in to her living quarters to oversee the birth, and the king, whose child it was, was said to have hidden behind the curtains during the birth process. (She was Louis XIV's mistress and bore him four children.) The mother's face was covered with a hood to offer a semblance of privacy. Clément was said to attend all of her subsequent births, and his attendance at important court births began to change the custom of who oversaw childbirth. (Two of the four children born to the duchess were eventually legitimated, which increased her standing in society.) Over time, more men, called *accoucheurs,* were appointed to jobs in *lying-in hospitals.* This was the beginning of men being chiefly in charge of the birthing process.

By the 18th century, more men were attending women in childbirth. The leading obstetrician in London was William Smellie (1697–1763). His well-known *Treatise on the Theory and Practice of Midwifery,* published in three volumes between 1752–64, contained a good explanation of the various stages of the labor process, and he wrote the first systematic discussion on the safe use of obstetrical forceps (see "The Chamberlen Family Secret" below), which saved countless lives. He also devised a method that helped turn a breech baby for an easier birth. Smellie placed midwifery on a sound scientific footing and helped to establish obstetrics as a recognized medical discipline. The well-regarded physician William Hunter was soon to follow.

THE CHAMBERLEN FAMILY SECRET

While William Smellie was the first to write of the use of forceps in childbirth, it was suspected at that time—and has since been proven—that a family by the name of Chamberlen had created

Forceps

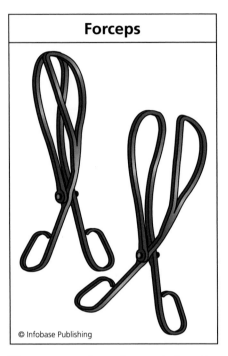

© Infobase Publishing

First created and used by the Chamberlen family, forceps greatly improved the odds of a safe birth for both mother and child. The family did not share their secret for many years, but, by the 18th century, other physicians were using the implement.

something similar that helped ease childbirth. The design of the instrument was kept secret and used only by family members, who served as midwives, for at least 100 years. Forceps were a revolutionary instrument that significantly reduced the mortality rate for women and their fetuses in difficult deliveries by changing the position of the fetus in the uterus to make the delivery safer and easier. Before the invention of forceps, difficult deliveries usually ended with an abortion of the baby or the death of both mother and fetus. If the mother died, then sometimes cesarean sections were successfully performed to save the baby.

While the idea that a family withheld this type of improvement from others is certainly not to their credit, the Chamberlens encountered resentment from the medical establishment for their stance that midwives should organize separately, which would have diminished the power of the College of Physicians. The antagonism between the college and the family may have left the brothers in no spirit to share their knowledge.

Peter Chamberlen (the elder) was the first Chamberlen who helped women through childbirth. He was born in Paris in 1560, the son of a Huguenot barber-surgeon. The family soon had to flee France for religious reasons, and they settled in Southampton where another son, also named Peter, was born. Both sons followed in their father's footsteps and became barber-surgeons, and they became known for their skills at helping with difficult births. Peter the elder moved to London in 1596 and became surgeon and accoucheur to Queen Anne, wife of James I, and he was soon joined by his brother. Peter the elder was eventually committed to prison for prescribing medicines that were contrary to the rules of the College of Physicians. The Lord Mayor interceded, and he was released. Then, the College of Physicians tried to have the younger Chamberlen put in prison to make up for the elder getting out, but their plan did not come to fruition.

Exactly which family member developed the obstetrical forceps and the year in which it was done are shrouded in mystery because the family carefully guarded their secret. Biographers write that the Chamberlens would arrive by carriage at the house of the expectant woman. They brought with them a wooden box so big that it had to be carried by two people. The box was adorned with gilded carvings, and it was rumored to contain a complicated machine that would help with childbirth. The laboring woman was blindfolded so that she could not see the secret device, and the Chamberlens requested that family members step out of the room before they set up for the birth process; then the door was locked. The Chamberlens were either showmen or people's imaginations ran wild, as those nearby reported that they heard ringing bells and other sinister sounds as the secret was put to work.

Peter the younger fathered another Peter who went into the family business and served Queen Henrietta Maria. His reputation was such that he was asked by the czar of Russia to attend a family birth; Charles I did not give permission. This Peter Chamberlen also worked to create a Corporation of London Midwives, but this move also met with great resistance from the College of Physicians and other midwives who decided they preferred remaining independent.

His son, Hugh, continued the family tradition. It seems that in 1670 he visited Paris, hoping to sell the family secret to the French government. François Mauriçeau, a well-respected obstetrician in France, asked Hugh to oversee the delivery of a baby of a dwarf with a very deformed pelvis, perhaps as some sort of test. Hugh was unable to provide effective help. Whether Hugh failed to make the sale to the French because he failed to save the dwarf or because Mauriçeau had no respect for him is unknown; some sources report that Mauriçeau accused Chamberlen of being a swindler and was aghast that the family would keep such an important invention a secret.

Chamberlen seemed to hold no ill will toward Mauriçeau, and he brought back a copy of Mauriçeau's very excellent book on obstetrics. In 1672, Hugh translated and arranged to publish it, including the following apology:

> My father, brothers and my self (tho' none else in Europe as I know) have, by God's blessing and our industry, attain'd to and long practis'd a way to deliver women in this case (obstructed labour), without any prejudice to them or their infants: tho' all others (being oblig'd for want of such an expedient to use the common way) do and must endanger, if not destroy one or both with hooks . . . I will now take leave to offer an apology for not publishing the secret I mention we have to extract children without hooks, where other artists use them, viz., there being my father and two brothers living that practise this art. I cannot esteem it my own to dispose of, nor publish it without injury to them.

The book, complete with his own apologetic explanation, brought Hugh to prominence, and he established a very successful practice and was appointed as a physician in ordinary to King Charles II. Hugh had no male heir, and obstetrician William Smellie believes that because of this Hugh let the secret leak out. Certainly by 1733, forceps were being used. The English physician Edmund Chapman wrote about the design and use of the obstetrical forceps, and historians feel his account accurately reflected the device that the Chamberlens used though there was no proof of this for a very long time. Then in 1813, the floorboards in the attic of the home in which Peter Chamberlen used to live were lifted and five pairs of obstetric forceps were revealed. His wife, Ann, had evidently hidden them at the time of his death, 130 years earlier. Reportedly, they were remarkably well designed and had a cranial curve for grasping the head. These instruments are now in the possession of the Royal College of Obstetricians and Gynaecologists in London.

WILLIAM HUNTER (1718–1783): NOTABLE OBSTETRICIAN

William Hunter was the first of two remarkable brothers who were to make important contributions to medicine. William began with a specialty in physiology and anatomy and ran a well-respected school of anatomy in London; he later devoted his practice to obstetrics.

William had intended to go into the ministry, but the well-respected physician and University of Glasgow professor William Cullen spotted William's abilities and encouraged him to go into medicine. A strong relationship developed between the two men, and Cullen invited William to join his practice to eventually take over the surgical duties. The two men agreed that Hunter should go to Edinburgh and London to further his medical knowledge. While away, the young man discovered the excitement of studying anatomy and, with Cullen's blessing, stayed in London to pursue his opportunities. He soon began lecturing on both anatomy and surgery

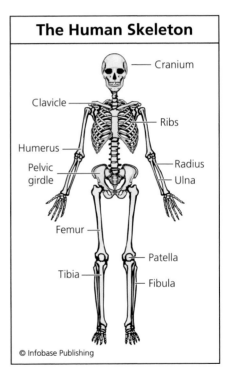

The Human Skeleton

Cranium
Clavicle
Ribs
Humerus
Pelvic girdle
Radius
Ulna
Femur
Patella
Tibia
Fibula

© Infobase Publishing

The study of anatomy was still not part of the curriculum at most medical schools, so those who wished to study the human body had to pay to attend classes at a separate school that specialized in these courses. William Hunter ran one such school of anatomy.

for an organization of navy surgeons. He went on to serve as surgeon-accoucheur at Middlesex Hospital (1748) and the British Lying-in Hospital (1749). To firm up his credentials, he returned to the University of Glasgow in 1750 to obtain his degree. (At that time, the degree of doctor of medicine was granted when the candidate presented certificates from other doctors of medicine that verified his qualifications and, most important, paid an agreed-upon sum for the degree.) In 1764, he was appointed physician to Queen Charlotte Sophia. When the Royal Academy finally added anatomy to their medical curriculum, they turned to William Hunter to serve as the first professor in that field.

Hunter's greatest work was *Anatomia uteri umani gravaidi* (The anatomy of the human gravid uterus exhibited in figures), published in 1774. Within Hunter's text were schematic illustrations similar to those created by the 15th-century artist, Leonardo da Vinci, showing the various stages of the developing embryo within the uterus.

Hunter's knowledge of anatomy made him of value within the field of natural history, too, and when French academicians returned from the United States with bones from what was presumed to be an elephant found along the banks of the Ohio River, Hunter concluded that it was not an elephant but an animal incognitum (probably the mammoth of Siberia).

William lived for his work. In 1783, he suffered a particularly severe case of gout that had kept him bedridden. One morning he felt a little better, so he decided to get up and give the introductory lecture on surgery at his school. By the end of the lecture, he was so exhausted he fainted and had to be carried out of the lecture hall by servants. A few days later, he died.

Hunter himself contributed greatly to the improvement in studies of anatomy as well as advancements in obstetrics, but he also developed pupils who had great influence in the field. William Shippen moved to the colonies and taught anatomy and midwifery in Philadelphia from 1763, and he helped establish the domination of male accoucheurs in North America. Shippen also introduced other changes. He believed women should give birth in rooms that were light and airy, and breast-feeding was encouraged. The light, airy rooms eventually made way for light, airy, and sanitary rooms, and the idea of breast-feeding was a return to the past that benefited both mother and baby.

JOHN HUNTER (1728–1793): BRITISH ANATOMIST AND SURGEON

John Hunter's background made it quite surprising that he emerged to be the preeminent surgeon of his day. The 10th and last child, John was much indulged. He attended only a few years of elementary school, obtaining no further education after these early years. Then, at the age of 20, he wrote to his older brother, William, and asked if he could come to London and assist at William's anatomy school. (Anatomy was not considered part of medical school so it was taught through private institutions; classes were small because not many students could pay for the additional education.)

William quickly agreed to his brother's visit, and, shortly after he arrived, John was given an arm from a cadaver and told to isolate the muscles. John excelled at this, and, after a second "test," he was taken in for further instruction by William and his assistant. Over time, John began to handle most of the dissections for the school, though his lack of education made him a poor lecturer. In 1749, John also began attending at the Chelsea hospital where he saw firsthand the nature of illness and injuries.

In the late 1750s, John Hunter's health began to suf-

John Hunter was one of the preeminent surgeons of his day; engraved by W. Holl *(National Library of Medicine)*

fer, and it was determined that he was spending too long in the autopsy rooms, so, in 1760, he pursued an appointment to be a staff surgeon in the military. England was involved in wars on several fronts at the time, and first John was sent to Belle Île, a small island off the coast of France, and later he was transferred to Portugal, where he helped with the wounded. His war-related experience was later reflected in his *Treatise on the Blood, Inflammation and Gun-Shot Wounds,* which was not published until 1794, the year after his death.

When he returned to London, he continued doing surgery and dissections, but, during his free time in the military, he had become fascinated with the animals and natural history of the areas where he was stationed. This fascination continued, and he acquired an assortment of animals to study (including leopards, jackals, goats, and rams), as well as ducks and geese so that he could gather their eggs to conduct embryological studies. While in the military, John had undertaken embryological studies of eels, so he also made arrangements with a fishmonger to bring in eels every month as he continued to attempt to identify the eels' ova-

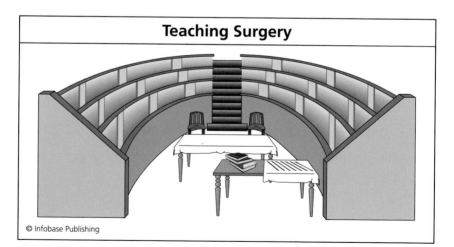

Teaching Surgery

© Infobase Publishing

Surgery was still usually taught by one physician giving a lecture while another person demonstrated what should be done using a cadaver.

ries. Because he had no more powerful tool than the naked eye for his dissections, he was never able to achieve this goal. Hunter helped found the Royal Veterinary College and did notable work on animals as diverse as whales and opossums.

In the winter of 1773, John became annoyed that others were teaching his surgical methods, sometimes incorrectly, so he decided to organize a lecture series to teach his systematic principles of surgery. Because lecturing made him so uncomfortable, his biographer, Everard Home, wrote that he had to take 30 drops of *laudanum* (opium) before each lecture.

Bodies were not easy to come by, and even esteemed physicians such as the Hunters had to resort to underhanded means to obtain bodies. (See "Procuring Bodies for Study" on page 34) In 1783, John Hunter was said to have bribed an undertaker £500 for a particularly noteworthy specimen—the corpse of an Irish giant who was so oversized that he had been a circus attraction. The giant had wished to be buried at sea, but Hunter bribed the right people, was able to conduct some study of the cadaver, and eventually put the skeleton on display.

By the end of his career, John Hunter had many accomplishments. He was first to use surgical ligation to correct an aneurysm (1785); wrote *Natural History of the Human Teeth* (two volumes, 1771 and 1778), which advanced dentistry; studied and wrote about digestion delivering a paper, *On the Digestion of the Stomach after Death,* to the Royal Society in 1772. He studied comparative anatomy, the lymphatic system, and examined inflammation and gunshot wounds. He also studied venereal disease. There are conflicting reports on how his study was conducted. Some say he injected diseased pus into a nephew and observed the fellow's reaction over time. Another report had it that he injected himself with infected pus and then had to delay his marriage until he had undergone a cure. (There actually was an unexpectedly long—three years—delay between his engagement and his marriage, but no one knows the reason.)

John was committed to improving medical education and instituted a medical society, Lyceum Medicum Londinense. In 1783, he purchased a home that was large enough to house his growing collection of specimens and also built rooms for conducting classes. By 1792, he had turned all lecturing duties over to Sir Everard Home so that he could perform surgery and write. The part of the home he dedicated to his collection became a comprehensive museum of comparative anatomy featuring items from Belle Île and Portugal as well as local specimens. At the time of his death, the museum contained some 14,000 preparations, most of which had been prepared by Hunter himself.

Biographer Everard Home eventually burned many of John Hunter's notes. Some biographers feel that his own work was heav-

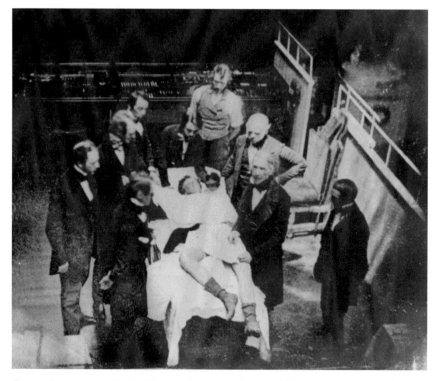

Operating room of the Massachusetts General Hospital, Boston. This is a reenactment of the first operation under anesthesia (ether), which took place on October 16, 1846. Daguerrotype by the famed Southworth & Hawes partnership of Boston *(Library of Congress Prints and Photographs Division)*

ily based on Hunter's work and therefore he could not afford for people to see Hunter's original notes.

THE EARLY USE OF ANESTHESIA

Prior to 1846 and the introduction of anesthesia, surgery was almost always limited to the extremities and superficial parts of the body. Anything else would have been too painful. However, the introduction of anesthesia did not lead to an immediate acceleration of surgical advances as there were still many other aspects of medicine that needed to be understood, such as how to control infection and pathology so physicians had some idea of what they were accomplishing surgically.

There were two forms of anesthesia that came into use in the 1840s: One was nitrous oxide (often called laughing gas) and the other was ether, which became the more popular for surgery. Nitrous oxide was first used with dental patients, and, over time, it became more commonly used by dentists than physicians. (Some dentists today still use it.) For surgical patients, ether was often administered by having a patient breathe into a cloth saturated with ether, and this method of drug delivery was used into the 20th century. Because physicians were in disbelief that surgery could be accomplished without caus-

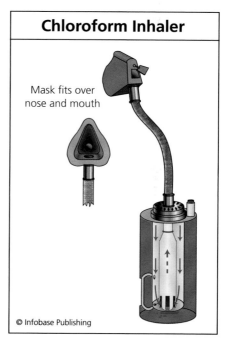

Chloroform Inhaler

Mask fits over nose and mouth

© Infobase Publishing

As physicians experimented with substances more efficient than alcohol to render a patient unconscious during surgery, one of the intoxicating agents tried was chloroform, administered by this type of inhaler and first used in the mid-19th century.

ing pain to the patient, these early operations with the patient under ether were frequently performed in front of audiences.

PROCURING BODIES FOR STUDY

While the days since Leonardo da Vinci had to study anatomy by candlelight in a well-hidden location so that no one would know he was doing dissection were in the past, things had only advanced a little. People understood the necessity of studying the human body, but that did not mean that cadavers were easy to come by. With a decrease in capital punishment, professors and students had to explore underground ways for getting bodies. As a result, a full-fledged profession developed for people who were willing to procure cadavers for pay. Calling themselves *resurrectionists* or resurrection-men, these fellows came up with several methods for obtaining bodies.

The poorhouses were valuable sources, so the resurrectionists developed contacts who would notify them when someone died. Other men employed women who could claim a body by arriving at a poorhouse acting the part of a grieving relative. Bribing servants to give up a master's body and putting stones into the coffin instead was also sometimes done.

Other resurrectionists trolled cemeteries for what they needed. In New York and Philadelphia, public officials and burial ground employees were routinely bribed for entrance to the *potter's fields* to get bodies. Graves of black people were more readily raided than those of white people, but white bodies were not safe, either. Resurrectionists would arrive in the dark of night. They dug quickly with a wooden spade to avoid the clanging sound of a metal one. They mastered the art of unearthing just one end of the coffin and then using a crowbar to open and break the top (the weight of the earth on the other end of the coffin lid helped them snap it off). A rope was then put around the body so it could be dragged out. Resurrectionists prided themselves on leaving clothing and jewelry behind; body-snatching was a *misdemeanor* while thievery of the belongings upgraded it to a *felony*.

THE DOCTORS' RIOT

At City Hospital in New York on April 13, 1788, a young boy peeking through the hospital window became concerned that the body the physicians were dissecting was that of his recently deceased mother. He reported this to his father, a mason, who led a group of laborers to the cemetery to check on the grave, and, when they opened it, the grave was empty. The group then moved on to attack the hospital. The statesmen John Jay and Alexander Hamilton arrived and tried to calm the mob, but they were stoned for their efforts. For their safety, the medical students were jailed.

Recently, a contemporary account of this event was found in a collection of papers belonging to John Marshall at the Institute of Early American History in Williamsburg, Virginia, and it was reprinted in the *Bulletin of the New York Academy of Medicine.* The letter, quoted below, was written by a colonel in the militia reporting to Governor Edmund Randolph on what was happening:

> We have been in a state of great tumult for a day or two past—The causes of which as well as I can digest them from various accounts, are as follows. The Young students of Physic, have for some time past, been loudly complained of, for their very frequent and wanton trespasses in the burial grounds of this City. The Corpse of a Young gentleman from the West Indias, was lately taken up—the grave left open, & the funeral clothing scatterd about . . . The cemeteries have been watched carefully. Then: On Sunday last, as some people were strolling by the hospital, they discovered a something hanging up at one of the windows, which excited their curiosity, and making use of a stick to satisfy that curiosity, part of a

(continues)

(continued)

> man's arm or leg tumbled out upon them. The cry of bar-
> barity &c was soon spread—the young sons of Galen fled
> in every direction—one took refuge up a chimney—the
> mob raisd—and the Hospital apartments were ransacked.
> The mob reassembled on Monday and more destruction
> followed.
>
> Despite these difficult circumstances, change did not
> really occur in this practice until the 1820s.

Body-snatching presented a terrible problem for the families of the deceased. They commonly set up watch over the body until burial and then someone would station themselves near the grave for a few days to be certain it was not dug up after burial. However, the body snatchers were quite artful; they sometimes tunneled into a recent grave after digging a hole 15–20 feet (4.6–6.1 m) away. The end of the coffin was then removed and the corpse was pulled out through the tunnel.

Medical students were often responsible for procuring their own bodies, and documents indicate that the procurement of bodies was actually quite stressful. One fellow wrote: "No occurrences in the course of my life have given me more trouble and anxiety than the procuring of subjects for dissection." With his friends at Harvard, this fellow, John Collins Warren, Jr., created a secret anatomic society in 1771 called Spunkers, whose purpose was to conduct anatomic dissections.

The first law that was somewhat helpful in delivering bodies for use by medical students was the Murder Act of 1752, and it stipulated that the corpses of executed murderers could be used for dissection. In 1810, a society was formed to begin to push for an alteration in the law since there were not enough bodies. The final straw came in the late 1820s when two resurrectionists, Burke and Hare, in Edinburgh decided the way to get truly fresh corpses, which commanded more money, was by murder. This and the "London Burkers" that followed (London Burkers were body snatchers who continued the practice of Burke and Hare and murdered victims so they would have bodies to sell to anatomists) stressed the importance of passing the Anatomy Act of 1832, which allowed that unclaimed bodies and those donated by relatives could be used for the study of anatomy. Anatomy teachers also needed to be licensed by this rule. All these new laws led to a decrease in the practice of body-snatching.

CONCLUSION

"Don't think, try the experiment," were favorite words of John Hunter, and that well explains the spirit of the times when it came to issues of anatomy, surgery, and childbirth. The physicians were pushing forward to learn more, and, while not all the things they learned provided them with helpful answers, the work they did certainly began to point them in a new and improved direction.

3

Changes in Battlefield Medicine

Throughout the 17th century, little thought was given to caring for the wounded on the battlefield. During earlier times when swords and arrows created most of the battle injuries, individuals took care of their own wounds or they might be helped by a fellow soldier. Most had to walk off the battlefield themselves as there was no method for transporting them at that time. Many of the badly wounded were simply left to die; some who were not fatally wounded but were not mobile could not help themselves and eventually died of hunger or thirst lying in the spot where they were injured. A doctor or barber-surgeon might be sought out later to provide additional aid, but this occurred only after the battle was over.

At this same time, medical knowledge in Europe was beginning to increase. Physicians now understood the circulatory and respiratory systems, the microscope was being used to learn more about the functions of the body, and discoveries in chemistry and physics were occurring with some regularity. Surgical instruments were greatly improved, and, with the invention of the printing press, the experience of other physicians could be written down and read about by others.

However, despite these advances, treatment of regular patients—as well as wounded soldiers—was about the same. The ligation method used by the Romans was still used to control bleeding from the arteries, and not until the 18th century was there progress in amputation. With the advent of deadlier ways to use gunpowder, more serious wounds were occurring and the results were devastating. Sometimes the wounds themselves were fatal because the bleeding was not stopped quickly enough; other times the gunpowder and other unclean projectiles would settle into the wound, causing deadly infections. Salves, ranging from all sorts of potions to dung, were placed on wounds, and many of the treatments were useless, some even harmful. Sometimes gunshot wounds were treated by applying a mixture to the soldier's weapon, and the concoctions were often so dangerous that sometimes this was preferable to putting these concoctions on the wound.

It became very clear that these methods of self-aid and inconsistent professional attention after battle were inadequate. Change was afoot. This chapter will explain the state of battlefield medicine in the early 1700s and introduce innovators like Dominique-Jean Larrey, the first modern military surgeon, who addressed getting treatment to the soldiers on the battlefield and helped bring about new treatments. Europe's armies saw many types of improvements as the century wore on, but few of these changes were possible for the ragtag group that made up the army during the Revolutionary War.

THE STATE OF BATTLEFIELD MEDICINE

By the early 1700s, the men on the battlefield were aware that certain medical improvements were vital if any of the soldiers were going to survive. Bandaging and wound-dressing became a skill that many learned. They applied pressure with sponges or sometimes applied *styptics* to stop minor bleeding. Once the soldiers got off the battlefield, physicians were devising new ways to help them. One of the devices created during this time was a surgical tool for removing musket balls. However, no one understood the importance of sterilization so the statistics on survival

Hippolyte Delaroche: *Napoléon Crossing the Alps*, Paris, 1848 *(Musée du Louvre, P.d.)*

showed little improvement. A French surgeon, Jean-Louis Petit (1674–1750), invented a screw tourniquet that was very helpful in controlling bleeding, and the device made thigh amputations possible. The screw tourniquet actually had quite a long life, and it was used for the next 160 years—until after the American Civil War.

Pierre-Joseph Desault (1744–95), a military surgeon, developed a technique for treating traumatic wounds by removing the dead tissue that usually removed the source of the infection. Percival Pott (1714–88), a British military surgeon, reduced the risk of infection in head wounds by developing a method for draining the wounds. As amputations became somewhat safer, military surgeons gave greater emphasis to preparing limbs for prosthetics. The death rate from amputation remained high until methods were developed in the 19th century to control infection and shock.

John Pringle (1707–82), the physician general to the British forces in 1740, identified jail fever, ship fever, and hospital fever as being one disease, now known as epidemic typhus. This discovery had a great impact on the military and will be fully discussed in chapter 4.

Though Pringle was only in active service for six years, the observations he made during this time were key to the world of medicine. A very simple request by Pringle made a big difference to the soldiers—Pringle asked that every man be issued a blanket, something that had not been done before.

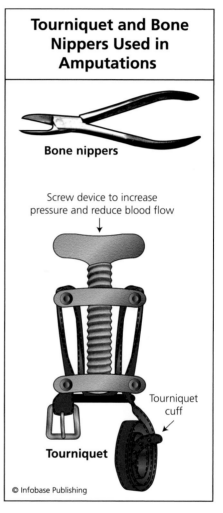

Tourniquet and Bone Nippers Used in Amputations

Bone nippers

Screw device to increase pressure and reduce blood flow

Tourniquet cuff

Tourniquet

© Infobase Publishing

Larrey made improvements in the methods used for amputations, and he also established a 24-hour rule. To be successful, amputations needed be done quickly, and no more than 24 hours should pass before the surgery was completed.

He also established that military hospitals needed to be recognized as a neutral territory so that medical help could be located nearer the battlefield without threat of attack.

Other physicians and surgeons also began to write about ways to keep armies healthy. In 1764, Richard Brockelsby (1722–97), an English physician, wrote a book on controlling contagious diseases in military hospitals. In 1794, the Scottish surgeon John Hunter's (see chapter 2) treatise on gunshot wounds was published. In it, Hunter argued against additional bloodletting after a gunshot wound.

THE FIRST MODERN MILITARY SURGEON

Dominique-Jean Larrey (1766–1842) can be credited with introducing field hospitals, ambulance service, and first-aid practices to the battlefield. He saved countless lives during the Napoleonic Wars and created a model for casualty transport that would serve armies well into the 20th century.

Dominique-Jean Larrey was born in a small village in the Pyrenees and was orphaned at the age of 13. The uncle who raised him was chief surgeon in Toulouse, and Larrey's career choice was dictated by the opportunity to apprentice under his uncle. Larrey's apprenticeship lasted six years, and he showed remarkable ability so he was sent to Paris to further his studies at the Hôtel-Dieu. Larrey shifted his plans when he got to Paris and joined the French navy instead. He became chief surgeon on a *frigate* traveling to North America. In 1792, he was in France when war broke out, and he became an assistant surgeon in the French army on the Rhine.

During one of Larrey's first experiences in battle (campaign of the Rhine, 1792), he noted the problems with caring for the wounded: "The wounded were left on the field, until after the engagement, and were then collected at a convenient spot, to which the ambulances speeded as soon as possible; but the number of wagons interposed between them and the Army and many other difficulties so retarded their progress that they never arrived in less than 24 or 36 hours, so most of the wounded died for want of assistance . . ." quotes Captain José M. Ortiz from Larrey's memoirs in an article on Larrey in the *U.S. Army Medical Department Journal.*

Larrey was not writing about an isolated case; the problem was widespread. In 1788, France had issued a royal ordinance requiring the creation of better transportation for the battle wounded, and, in 1792, a National Convention was set up to discuss it. In 1793, representatives from the commission offered a prize for the transportation design that best fit the commission's specifications. However, after eight months, the commission had not yet seen a design they were satisfied with.

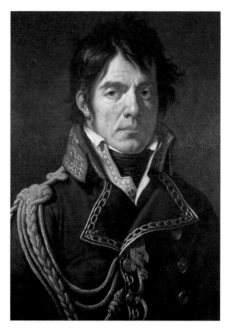

Dominique-Jean Larrey by Anne-Louis Girodet-Trioson *(The Yorck Project)*

Those on the battlefield had to take a more practical approach, and so Larrey simply set about looking for a solution. In 1797, he was appointed to aid in the medical affairs of the military campaign in Italy, and this gave him the opportunity to test a theory he felt would work. During his time on the battlefield, Larrey had observed an operation—a horse artillery—that the French army had perfected to a high degree. First used by the Swedish during the Thirty Years' War, this artillery unit featured mounted soldiers on horseback who could mobilize quickly to swoop in and bring additional firepower to supplement the infantry wherever needed. By the 18th century, King Frederick the Great adopted the method for use in Prussia in 1759, and he insisted that his men drill relentlessly so that they understood the importance of

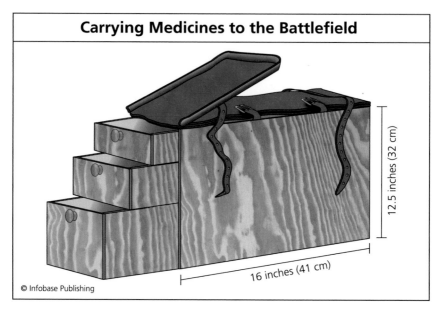

Carrying Medicines to the Battlefield

12.5 inches (32 cm)

16 inches (41 cm)

© Infobase Publishing

Larrey realized that to work efficiently on the battlefield, physicians needed to be well organized.

mobility and speed in their operations to bring in more power-ful firepower quickly. Soon other armies were using the system, but the largest and most efficient system was organized by the French revolutionary army. Larrey had plenty of opportunity to observe the speed and efficiency by which the well-practiced, mounted crew could operate. He thought if he could introduce a similar system to aid battlefield medical care, it would really make a difference.

He realized that a well-organized casualty transport system was key. His rank permitted him to gather a legion of 340 men, made up of officers, sub-officers, and privates. He broke the unit into three groups of 113 men, each commanded by a chief sur-geon, and arranged for each division to have 12 light and four heavy carriages; each carriage was manned by a crew of seven. Everyone had a specific assignment. One fellow was the bearer of surgical instruments, another carried surgical dressings; 25 foot soldiers accompanied the group to care for the wounded.

According to Captain Ortiz's article, it is known from Larrey's memoirs what the various soldiers were assigned to carry: each surgeon "carried a small *cartouche* box . . . divided into several compartments, containing a case of portable surgical instruments, some medicines, and articles necessary for affording immediate assistance to the wounded, on the field of battle. . . ." Officers were also given courier bags that attached to the saddles that contained field dressings (this was instead of pistol holsters). Others were there to solve other types of problems. There was a *farrier* (blacksmith), a saddler, and a bootmaker to help get the men moving again.

Each division had 12 light carriages; some had two wheels for use on level ground; some had four wheels for getting through more rugged terrain. The ambulances traveled in a set order, and procedures were outlined for what to do with the dead. Larrey described his basic ambulance as follows: "The frame . . . resembled an elongated cube, curved on the top: it had two small windows on each side, a folding door opened before and behind. The floor of the body was moveable; and on it were placed a hair mattress, and a bolster of the same, covered with leather. This floor moved easily on the sides of the body by means of four small rollers; on the sides were four iron handles through which the sashes of the soldiers were passed, while putting the wounded on the sliding floor. These sashes served instead of litters for carrying the wounded; they were dressed on these floors when the weather did not permit them to be dressed on the ground.

The larger ambulances were drawn by four horses and had two drivers. They also had a compartment in the back to carry food for the horses and could carry four men with some effort. The window allowed for ventilation. Compartments on the sides of the carriage provided storage for medicine and materials for bandaging. In addition, items such as a handcart could be attached to the carriage.

Adaptations were always necessary. In rugged areas, mules were needed for carrying supplies as the carriages could not be weighed down. In the deserts of Egypt, camels were used to help with transport, and camels mostly replaced the other pack animals as well since they were more useful in the desert.

Larrey had planned the ambulance corps so that they could follow the most rapid guard within the army. The corps could also separate into smaller divisions, and since every medical officer was mounted and commanded a carriage, they were quickly able to reach the wounded with all the necessary supplies. In 1799, in Egypt where Larrey's flying ambulance corps operated, it was reported that none of the injured were left for more than a quarter of an hour without being dressed. Larrey's ambulances were commanded with the control and coordination of a seasoned field commander and their speed and flexibility permitted them to travel where needed.

Larrey was not alone in experimenting with ways to get injured men off the battlefield. Another French military physician Baron F. P. Percy also introduced a casualty transport system. Percy's method worked from the concept of being a mobile hospital. Medical professionals were transported to an area near a battle, and litter bearers went out to pick up the wounded and bring them to the chosen location. Larrey's method involved providing initial treatment before transport, which ultimately was more effective,

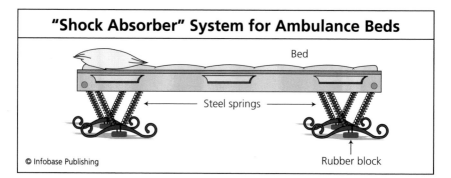

"Shock Absorber" System for Ambulance Beds

Bed

Steel springs

Rubber block

© Infobase Publishing

The earliest ambulance did not have anything as elaborate as shock absorbers. By the American Civil War, they had begun to explore ways to make travel more comfortable for the patients. Ambulances with these types of cots would have been rare and were probably used only for injured officers.

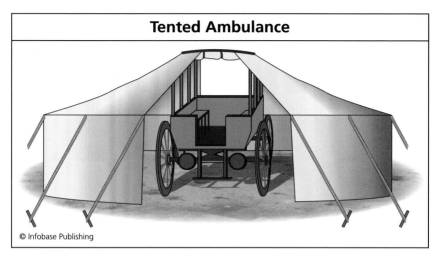

Tented Ambulance

© Infobase Publishing

Because travel was slow, this type of tented ambulance provided a way to travel greater distances with a patient.

but Percy was the first to introduce a "regularly trained corps of field litter bearers, soldiers regularly formed and equipped for the duty of picking up the wounded and carrying them on stretchers to the place where means of surgical aid were provided." This was a dangerous and physically strenuous assignment for the men involved, but also militarily significant because it relieved the common infantryman of the arduous task of caring for their wounded.

Both men created specialized ambulance corps and greater mobility for their physicians, and the guidelines they used to evacuate men were the basis on which military systems were later developed in Korea and Vietnam with helicopters.

LARREY'S OTHER ACCOMPLISHMENTS

As the French began to improve their survival rates, the English were still faring poorly. Experts debate how much this had to do with the flying ambulance and how much this had to do with the superior surgical skills of the French physicians. Larrey was known as a great commander as well as a great physician. He participated in 25 campaigns and 60 battles, and in 1801 Larrey became the surgeon general of the Imperial Guard, Napoléon's elite personal reserve corps.

THE IMPORTANCE OF TRIAGE

Triage is a method used to sort out patients by the severity of their conditions. This permits health care workers to help as many people as possible by providing the care that is most important immediately. While most sources report that the first organized use of triage occurred in World War I and was implemented by French doctors, this is discounting the fact that Dominique-Jean Larrey was the first to introduce a system of dispensing care that was not based on military rank. Larrey also established a rule for the triage (from the Old French word meaning "to sort") of war casualties, treating the wounded according to the seriousness of their injuries and urgency for need of medical care, regardless of their rank or nationality. Enemy soldiers were treated just as the French and their allies were.

In general, triage divides victims into the following three basic categories:

1. Those who are likely to live, regardless of what care they receive;
2. Those are likely to die no matter what is done for them;
3. Those for whom immediate care will make a difference.

During Larrey's time, the level of medical expertise was still not very advanced, but men could make an educated guess based on blood loss and number and location of the wounds a soldier had received. Implementing this type of method was a huge improvement over commanders who had minor

During his time in the military, Larrey made many notable observations. He realized that when a limb is badly damaged in battle, the speed of amputation can make the difference between

injuries being treated before enlisted men who were more seriously—but not fatally—wounded.

In today's military, they have refined triage so that very initial emergency care is conducted as quickly as possible but then patients are rapidly sent on to another station where they can be cared for more extensively. The frontline team may also specify which patients are moved out first and, if there is a choice of destinations, they will note where each patient is to be sent based on each person's particular needs. This organized system of interim medical care permits the emergency workers on the front line to address more of the injured in a shorter span of time.

After the initial triage at a disaster or on a battlefield, the next level of medical personnel makes a more detailed examination and also classifies the patients in the order that they should be helped. Ethical issues arise during this process. In an emergency situation—often with inadequate supplies—treating someone who is unlikely to live takes time and resources away from someone who may make it, yet making the decision that someone is past all hope places medical personnel in a quandary.

Today's lay and emergency personnel are trained to use a S.T.A.R.T. (Simple Triage and Rapid Treatment) model of managing major accidents or natural disasters. This system divides people into four groups: the deceased; the injured who can be helped by immediate transportation; the injured whose transport can be delayed; and those with minor injuries who need help but their needs are less urgent.

life and death, and this, to a large degree, was what drove his creation of the ambulance system. A few years later, he established a 24-hour principle that set the standard that an amputation should occur within 24 hours of a limb being shattered. Larrey was also the first surgeon to successfully amputate a leg at the hip, and he made many contributions in the treatment of leg fractures.

Larrey was a fellow who achieved recognition during his lifetime. Larrey was popular with the men because the very sight of the flying ambulance corps provided the men with hope—the first time their personal fears had been addressed. When Napoléon's troops were fleeing Russia over the last bridge crossing the Berezina River in 1812, Baron Larrey was specially lifted overhead by the crowd of troops so that he could get safely across before the other men.

At Waterloo, the duke of Wellington noted Larrey and ordered his soldiers not to fire in his direction to give him time to gather up the wounded. Napoléon himself was a fan of Larrey and once commented: "If the army ever erects a monument to express its gratitude, it should do so in honor of Larrey."

IMPROVEMENTS IN MILITARY MEDICINE

At the beginning of the century, the pattern of military medical care remained essentially as it had been in the previous century. By midcentury, however, all major European governments had moved toward providing food and medical care to all soldiers who were serving in the military. This included whatever was needed to maintain the health of the troops.

Food, Shelter, and Uniforms

Most soldiers who entered the military were coming out of poor living circumstances, so any regularly provided meals meant an improvement in their stamina. By the mid-18th century, governments were beginning to pay closer attention to providing enough food regularly, but the quality and quantity of food was often less than promised. Sometimes the government contracts were manipulated by fraud; other times, pressure to reduce government expenses affected the quantity or quality of food purchased for troops in the field.

In addition to providing the men with regular meals, the military began building barracks that were designed to house the troops when they were at their home base. (Before this, armies used to rent out inns or place troops in homes of residents.) The first British barracks were introduced in Ireland in 1713 due to a shortage of barns and inns.

New clothing for the soldiers was a worthy investment. Health practitioners soon learned that those

642   AMPUTATION OF THE LEG.

up for two inches, and turned back; and the muscles are to be divided down to the bone by a second circular incision. Then a long slender double-edged knife, called a catline, is passed between the bones to divide the inter-

Fig. 403.

[Flap amputation of the leg.]

osseous ligament and muscles, and both bones are sawn through together, the flesh being protected by a retractor, which should have three tails. The spine of the tibia, if it projects much, may be removed with a fine saw or bone nippers, and care should be taken not to leave the fibula longer than the tibia, or it will give much trouble. The anterior and posterior tibial and peroneal arteries, and any others requiring it, being tied, the stump is to be treated as directed after amputation of the thigh. The teguments should be put together transversely.

Fig. 404.

[Teale's amputation.]

4. *Teale's Operation.*—The length and breadth of the anterior flap are to be determined and marked out as before described, p. 635. The two lateral incisions are first made through the skin; and the transverse one down to the bone; then the long flap is dissected up, and with it, all the tissues in front of the bones and interosseous membrane; including the anterior tibial vessels, which are divided once only. The short posterior flap is then made by one cut down to the bones, and is to be dissected clean from the bones and interosseous membrane up to the point of sawing. The cuts are reduced from Mr. Teale's book.

A leg amputation from *Principles and Practice of Modern Surgery* (1860) by Robert Druitt (American Civil War Surgical Antiques)

entering the military needed to be bathed, and, because of lice infestations, it was a good idea to simply burn the clothing they were wearing on arrival. New clothing provided something for the men to wear, and a preplanned color provided easy identification on the battlefield. Because battles at that time were much more confrontational, ready identification was considered a plus until the Americans instituted the ambush style of attack where being able to lurk in the background was advantageous.

Unfortunately for the men, the uniforms were designed for the benefit of the country paying the bill. The priorities for selecting uniforms were affordability and making the men identifiable. The companies that made the uniforms frequently worked with cotton. The price for cotton was low, but the cloth provided little warmth and no protection against cold or rain. Tight buttons and belts often restricted breathing. Many units were assigned to wear tight stockings, which restricted the blood flow and provided insufficient padding to the bottoms of the soldiers' feet. The shoes themselves were not created for miles and miles of walking, and they provided little protection from frostbite and trench foot. The headgear selected was sometimes heavy, and none provided protection from shell fragments and bullets.

Better Health

The custom of putting new soldiers through a physical examination resulted from the fact that governments began noticing that those who joined the military tended to be underfed and often bug-infested. At first, each company commander was ordered to give a quick examination to each new recruit. In 1726, the French army began assigning a physician to conduct the examinations, and, approximately 40 years later (1764), they began assessing the recruits for physical fitness. The Prussian government instituted regular physical examinations of all soldiers in 1788, and, in 1790, the British army finally fell in line with what the other countries were doing and added mandatory medical examinations for those entering the British army.

As previously discussed, few armies organized transport systems to move the wounded off the field. It often took several days for the injured to reach the closest hospital (usually a nearby house or barn commandeered for this purpose), and it was not unusual for a third of the patients to die in transit from the front to the rear hospitals. More and more armies were beginning to organize mobile field hospitals that could be located near the battles, but few had enough staff to adequately address the injuries that were occurring on the battlefield. Military hospitals remained unsanitary, and disease continued to be the major threat to military manpower.

In 1743, after the Battle of Dettingen, the last time a British monarch personally led his men into battle, an agreement was made to declare medical personnel noncombatants and to give wounded enemy soldiers medi-

Field Hospital after the Battle of June 27— Savage Station, Virginia, June 30, 1862 *(The Civil War Home Page)*

cal treatment and return them after they recovered from their injuries. This created a need for more medical personnel in order that there would be enough staff to treat soldiers from both sides of the battlefield.

MEDICAL CARE DURING THE REVOLUTIONARY WAR

Whatever gains might have been made on behalf of soldiers by the European countries, these advances were totally absent for the motley assortment of men who gathered to form the American army to fight in the Revolution. The men were not well fed, and no one had any understanding of germs or the spread of disease. They lived in close quarters and suffered malnutrition and fatigue. During this period, more Americans died from illness than died in combat.

The military attempted to regulate cleanliness of the camps and provide bedding and a balanced diet, but it was far from ideal. In fact, soldiers often went weeks without changing their clothes. Diseases ran through the camps at a rapid pace.

On the American side, anyone with medical knowledge was pressed into service to help tend to the injured or the sick. Each regiment brought its own physician, but these hometown doctors varied in ability. Only a handful had graduated from the

Florence Nightingale was a pioneer in the field of nursing. Her observations of the care of the wounded during the Crimean War led her to campaign for better treatment for patients. *(Photo by Perry Pictures, Library of Congress Prints and Photographs Division)*

10-year-old Philadelphia Medical College, and another group—fewer than 300—were mainly graduates of European medical schools where admission requirements included a knowledge of the classics and enough money to pay for the degree, which was heavy on theory and light on any clinical training.

Besides caring for those wounded in battle, the camp surgeon was responsible for caring for the camp's diseased soldiers. The camp surgeon was constantly on the alert for unsanitary conditions in camp that might lead to disease. Common diseases suffered by soldiers were dysentery, fever, and smallpox, brought about by bad sanitation. Hospitals were set up temporarily, usually in a local home near the camp. If soldiers were sent on to an official hospital, they were often overcrowded, lacking in supplies and cleanliness, which increased the death rate.

The most common type of surgery was the removal of musket balls from wounds or the bandaging of stabs from bayonets. In cases where the bone was damaged so severely that a limb could not be saved, the surgeon performed an amputation without anesthesia or any type of sterilization. Before an amputation, officers were generally offered rum or brandy to numb the pain, but enlisted men did without. Two fellow soldiers or two medical personnel would hold the patient on the table, and a tourniquet was placed four fingers above the line where the limb was to be removed. Then the surgeon used his amputation knife to cut down to the bone of the damaged limb. Arteries were moved aside by tacking them away from the main area with crooked needles. The surgeon used a bone saw; a small one was used on arms, a bigger one to remove a leg above the knee. A good surgeon could make the cut in about 45 seconds. Then arteries were buried in tissue skin flapped over and sutured. The stump was bandaged with linen, and the patient, whose temperature generally plummeted and went into shock, was stabilized when possible. Only 35 percent of the people who went through this procedure survived.

The Continental Congress created the hospital department for the army. The original department consisted of administrators and a corps of physicians for the Continental army. The army physicians did not wear uniforms until 1816, and they were not given military rank until 1847. Over time this department began to establish acceptable treatments of injuries and illnesses and a formalized list of qualifications for physicians.

Dr. Benjamin Church (1734–78) is a name that frequently surfaces in writings about the Revolutionary War. He was an active member of the Sons of Liberty and counted John Adams, Samuel Adams, John Hancock, and Paul Revere among his colleagues. He was the first physician on the scene at the Boston Massacre of 1770 and tended to the wounded and dying. He was a well-respected member of the Boston Committee of

Dr. Benjamin Church *(National Library of Medicine)*

Correspondence that helped push for the Revolution and, because he was a physician, he was appointed to be the director of the Continental army hospital in Cambridge in 1775.

Paul Revere was one of the first to be suspicions of Church. When information from a secret meeting Church and Revere had attended was leaked to British officers, Revere became concerned, but it took several more years before Church's status as a spy was revealed. Church had crossed military lines, saying he had to get more medical supplies, and he claimed to have been detained and questioned by General Thomas Gage while on the other side. Over time, word leaked out that Church had actually sought out Gage, and later he was caught sending coded messages to a British naval commander. The Americans court-martialed him, and in 1778 he was deported to the West Indies. His ship was lost at sea.

Other Challenges of the Battlefield

Concern about air quality during battle is not a modern problem. As early as the Revolutionary War in the United States, soldiers were plagued by bad air that made it difficult to breathe. In the 1700s, great amounts of dust and dirt were stirred up by the troops

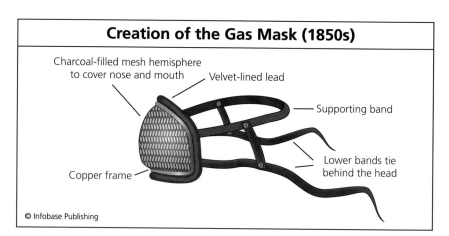

Creation of the Gas Mask (1850s)

Charcoal-filled mesh hemisphere to cover nose and mouth

Velvet-lined lead

Supporting band

Copper frame

Lower bands tie behind the head

© Infobase Publishing

Though earlier devices had been created to reduce the amount of smoke or noxious air breathed in by soldiers or firefighters, this device was created in 1854 by Scottish chemist John Stenhouse. It used charcoal to filter smoke or noxious gases.

JOSEPH LISTER (1827–1912):
Develops Method of Antiseptic Surgery

By the middle of the 1800s, postoperative sepsis infections accounted for the death of almost half of the patients undergoing major surgery. A chemist by the name of Justin von Liebig determined that sepsis occurred when the injury was exposed to air, so starting in 1839, he advocated that wounds should be covered with plasters.

British surgeon Joseph Lister was skeptical of this explanation and von Liebig's recommendation. Lister had devoted a good number of years to studying inflammation of wounds at the Glasgow infirmary, and went on to eventually be the surgeon in charge of the Glasgow Royal Infirmary. During his work, Lister noted that 45 to 50 percent of the amputation cases in the male accident ward were dying of sepsis (1861–65).

Lister suspected that a cleaner environment might be helpful. He began wearing clean clothes when he went in to perform surgery. (This was not the norm for the day; surgeons frequently considered it a badge of honor to appear in a blood-spattered apron.) He also washed his hands before each procedure. At first, Lister made no noticeable progress.

Then he became aware of work being done by Louis Pasteur, a chemist who was later to make great strides in understanding the cause of disease. Pasteur's work suggested that decay came from living organisms that affected human tissues, and Pasteur advocated the use of heat or chemicals to destroy the microorganisms. Lister determined that Pasteur's microorganisms might be causing the gangrene that so often plagued surgery patients and decided that chemicals would be the best way to stem the spread of microorganisms during and after surgery. He read that carbolic acid was being

(continues)

(continued)

used to treat sewage in some places, so he created a solution of carbolic acid and began to spray surgical tools, surfaces, and even surgical incisions with his newly created mixture. For the next nine months, his patients at the Glasgow Royal Infirmary remained clear of sepsis.

At first, London and the United States resisted this theory; though they quibbled less about the theory of germs, they disagreed with the use of carbolic acid. To overcome this resistance, Lister arrived to become chair of clinical surgery at King's College where he began performing surgery under antiseptic circumstances, and, without much delay, his methods were accepted. Within just a few years, other surgeons began using Lister's antiseptic methods, and, in 1878, Robert Koch demonstrated that steam could be used for sterilizing surgical tools and dressings. (Koch was to go on to make many other discoveries.)

While the methods of sterilization have changed over the years, the concept of antiseptic surgery is still vital to success in these procedures.

marching through the countryside, and cannons and muskets emitted noxious smoke. Battles would end with the air so dark that it was difficult to see, let alone to breathe. During the early 19th century, several different inventors worked on devising a "gas" mask that could be used for military purposes. Early masks were simple pieces of fabric that covered the mouth and nose with the hope of screening out harmful materials from the air. Then in the early 1800s, chemists realized that charcoal could be used to remove bad odors. A thin layer of charcoal was sometimes sprinkled over decaying meat, and this lessened the bad smell. Over time, several scientists began working with a way to use the charcoal in a gas mask, a technique that Scottish chemist John Stenhouse perfected in 1858, before the Civil War. This type of mask also was soon used in various industries where workers needed masks to filter out some of the pollutants.

CONCLUSION

The changes that occurred in medical care on the battlefield were of vital importance to the soldiers, and the new ideas and inventions that were tested on the battlefield proved helpful in medical treatment for the general population. From the necessity of treating the wounded soon after an injury to the process of sterilization, these new methods were to have a lasting impact on the world of medicine.

4

Curtailing the Spread of Disease

Cholera, plague, smallpox, measles, scarlet fever, malaria, typhus fever, typhoid fever, influenza, and, probably, gonorrhea and leprosy were among the diseases that circulated during the 18th and early 19th centuries. At the time, physicians were baffled by how illness traveled. The theory of *miasma* was popular, but there were other ideas as well. The Italian physician and scholar Girolamo Fracastoro (who had written an epochal description of syphilis, giving the disease its name) published *On Contagions and the Cure of Contagious Diseases* in 1546. Fracastoro proposed that many diseases are caused by transmissible, self-propagating, disease-specific agents, which propagate themselves in tissues of the infected host and caused disease. Fracastoro also suggested that illness could spread by direct contact (person-to-person) or indirectly (by touching surfaces previously touched by the ill). But for 150 years, Fracastoro's ideas on this subject had been ignored, and no one picked up on it now.

Some countries were beginning to be more interested in studying the pattern of disease, knowing that that could be helpful. By early in the 15th century, the Italian boards of health instituted a system of death registration, first for contagious diseases and subsequently for all diseases. In London in the 17th century, they started maintaining these types of records as well. These were definite steps forward, as was the spotting of "little animalcules" by the Dutch cloth merchant Antoni van Leeuwenhoek, but no one was pursuing the idea that minute organisms might be causing disease.

Nonetheless, they were highly motivated to try to find ways to halt the spread of illness. Diseases like smallpox, yellow fever, and malaria often struck communities with a vengeance, and everyone wanted to find ways to end the spread of disease.

This chapter examines the work of Sir John Pringle that may have helped pave the way for the work of Louis Pasteur, and it discusses how the people of the day came to understand what to do about typhus, smallpox, and scurvy. The fact that no one understood the nature of what caused illness makes their progress all the more remarkable.

SIR JOHN PRINGLE (1707–1782): A WISE OBSERVER

The Scotsman Sir John Pringle is always mentioned in passing as a "founder of military medicine," but his contributions are worth more than passing attention. He lived and worked in a world that did not yet know about bacteria and disease, yet Pringle's work was one of the contributing factors that helped pave the way for Louis Pasteur and his development of "germ theory."

Pringle was the son of a baronet and intended a mercantile career. However, he happened to hear the well-known Herman Boerhaave give a lecture on medicine, and he was hooked. He took up the study of medicine and began his practice in Edinburgh. In 1742, he was appointed physician to the British army, and three years later he became physician-general of the army.

In 1752, after he had returned to private practice, there was a serious outbreak of what was called jail fever, which caused the deaths of prisoners. Some of the prisoners had been brought to court, and they were so contagious that several judges as well as the Lord Mayor became ill and died. This spurred Pringle to write to Richard Mead about "The Hospital and Jayl-Fevers," in which he noted that diseases that spread in hospitals and that circulated in jails were related forms of the same illness, and he identified it as typhus. He traced the illness to soldiers who had fought in the Battle of Culloden, who then transmitted the illness to English troops who soon fell ill as well.

Pringle saw that people living in close quarters—hospitals and jails—caused infections to spread. Based on his observations, Pringle wrote *Observations on the Diseases of the Army* (1752), the first English text on military medicine, which could also be applied to other environments. He offered sound advice on preventing infections in places like hospitals, addressing the problems of hospital ventilation and camp sanitation by advancing rules for proper drainage, adequate latrines, and the avoidance of setting up camps near marshes. He suggested that military hospitals be treated as sanctuaries, mutually protected by everyone.

Though no one including Pringle understood how typhus spread Pringle made specific suggestions that he thought might reduce this disease that decimated military units. He recommended that when men entered the military their clothing should be burned, and new clothing should be provided at public expense. He also suggested that prisoners should be bathed and put in clean clothes before a court appearance. Regular washing of bedding and clothing were also recommended. All these suggestions would have been helpful.

Typhoid fever and typhus are frequently confused. Both were prevalent at this time, but they are two distinct illnesses that travel in very different ways. While typhus travels via lice that spread the disease to humans, typhoid fever comes from ingestion of food or water contaminated with feces from an infected person. (See the following sidebar "Typhus: What It Is and How It Spreads" on page 64) Like typhus, typhoid fever was difficult to avoid when troops were traveling from place to place and setting up temporary camps.

Pringle's ideas were well received, and James Lind, another Edinburgh physician discussed later in this chapter, began to introduce these ideas to the Royal Navy, and Richard Brocklesby took the ideas to the army.

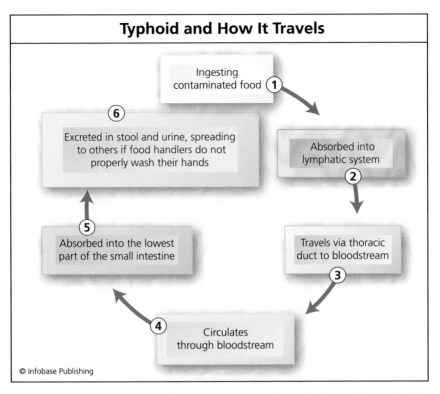

Typhoid and How It Travels

1. Ingesting contaminated food
2. Absorbed into lymphatic system
3. Travels via thoracic duct to bloodstream
4. Circulates through bloodstream
5. Absorbed into the lowest part of the small intestine
6. Excreted in stool and urine, spreading to others if food handlers do not properly wash their hands

© Infobase Publishing

Both typhus and typhoid fever were prevalent in the 18th and 19th centuries. Typhus was transmitted by lice and fleas. Typhoid (as depicted in the illustration) was transmitted via contaminated food.

As he continued to work and observe, Pringle broke rank with the majority of people who felt that there was an "Epidemick Constitution" (meaning that certain people were predisposed to become ill) and that miasma was the cause of illness. Pringle's fourth edition of his book describes how he thought people become ill, and he puts forward the possibility that those who were beginning to think that "animalcula" caused illness might want to further investigate it.

His later edition of the book also recommended the application of antiseptics (the use of strong acids) for cleaning surgical areas. Joseph Lister, another Scotsman (see chapter 3), was to later make more progress in this area. (It is not clear whether he knew of Pringle's work.)

JAMES LIND SOLVES THE PROBLEM OF SCURVY

Many health problems plagued people of this era, but not all of them required medicinal cures. This was before any understanding of the necessity of a balanced diet or the value of certain

---

TYPHUS:
What It Is and How It Spreads

During the 19th century, typhus spread during many of the wars that were fought and killed many of the soldiers. Ireland experienced a major epidemic in 1816–1819, in the late 1830s, during the 1840s, and during the Irish famine.

It is now known that typhus is a disease that comes from bacteria borne by lice. The lice live on mice and rats and transfer to humans, usually those who are living in overcrowded environments where it is often unclean. Typhus is also referred to as jail fever, hospital fever, ship fever, and famine fever, all appropriate labels. The illness begins with a severe headache and high fever, accompanied by a cough, severe muscle pain, falling blood pressure, sensitivity to light, and often stupor or delirium. A rash appears on the chest about five days after the fever begins.

In America, typhus epidemics occurred in Philadelphia in 1837 and in Baltimore, Memphis, and Washington, D.C. between 1865 and 1873. An epidemic in Concord, New Hampshire, killed the son of Franklin Pierce (14th president of the United States). Typhus also was present during the Civil War though typhoid fever (a very different illness, caused by the ingestion of contaminated food) was a bigger concern at that time.

Today, the disease can be treated with antibiotics, though additional fluids and oxygen are often needed to bring a person back to good health. The connection between lice and the spread of typhus was not found until 1909, at which point a *vaccine* was developed that has helped prevent infection.

vitamins (nutrition is discussed more fully in chapter 7), so there was little understanding about the importance of what one ate. This became a particular problem during the 17th and 18th centuries when explorers started one- and two-year journeys to explore other lands. Though it was very difficult to travel with adequate food supplies, the tradition was that a slightly better store of food was set aside for the captain. On long journeys, the captain often found that his crew became useless, suffering from stomach distress, blotchy skin, and bleeding gums that led to tooth loss. It was frequently fatal. (During the potato famine, the Irish also suffered scurvy, as potatoes provided some nutrients that helped guard against the disease.)

As early as 1614, the physician John Woodall, the surgeon general of the East India Company, wrote a book that noted that scurvy was a result of a dietary deficiency, and he recommended fresh food, including fruits like oranges and lemons.

However, it was not until 1747 that James Lind, a Scottish naval surgeon who came to be referred to as "the father of naval hygiene," formally proved that scurvy could be treated and prevented by supplementing the diet with citrus fruits. Lind ran what is considered an early clinical trial that proved that lemon juice could prevent scurvy. Lind divided a dozen scurvy sufferers into six groups of two, treating each pair with a different remedy for 14 days. The sailors given two oranges and a lemon each day recovered best. In 1753, Lind published *Treatise of the Scurvy,* and decades later the British navy adopted Lind's advice; scurvy was more or less eliminated.

The first explorer to circumnavigate the globe without losing a single man was Captain James Cook (1728–79). He took great quantities of pickled sauerkraut that could be stored better than fresh fruit and maintained a fair degree of ascorbic acid. Whenever they neared land, Cook also made a point of sending his men off in search of fresh food to supplement their diets.

THE DILEMMA OF SMALLPOX

Smallpox had existed for centuries. Smallpox may have emerged as early as 10,000 B.C.E., although it is difficult to assess what occurred prehistory. It went through differing waves of virulence; sometimes it was a mild illness known as variola minor; at other times it became more severe and few recovered—that strain was referred to as variola major. Epidemics frequently flared in areas as diverse as Asia, North Africa, Europe, and eventually America, where it was particularly devastating because the natives of North and South America had never experienced the disease so no one had immunity. In the late 1400s in Europe, syphilis began to appear and was known as the Great Pox, so the term *smallpox* came to distinguish the illness that had long plagued humanity.

The symptoms of smallpox appear about a week or two after a person is exposed to an infected person or a contaminated object. The victim develops a high fever and suffers from a pustulous full body rash. Those who survived the illness gained lifelong immunity, but smallpox killed 30–50 percent of those who became ill with it.

During the 16th century, the smallpox strain in Europe was a mild one, and people frequently had smallpox but rarely died of the disease. Then in the 17th century, a new strain emerged, and epidemics began to kill hundreds and thousands of people. Physicians had many different theories about it, but nothing was curative and nothing halted the spread of the illness. At this time, one in three who contracted the disease died, and those who survived were often badly disfigured. In addition, this was the age of exploration, so as explorers expanded their range and trade routes extended their paths, smallpox traveled along with the explorers, traders, and soldiers. The virus was very capable of maintaining its strength; it traveled via human carriers but also was spread by clothing and objects touched by those who were ill. During the 18th century, the disease killed an estimated 400,000 Europeans each year (including five reigning monarchs) and was responsible for a third of all blindness. Over 80 percent of children who became ill died, and among the adult population smallpox killed between 20 and 60 percent of those who became ill.

EFFORTS TO PREVENT SMALLPOX

The Chinese may have been the first (960 C.E.) to use scrapings from smallpox scabs or pustules to inoculate people so they would not get ill. This method was eventually used in Turkey and North and West Africa. One of the early Westerners to note the practice was Lady Mary Wortley Montagu (1689–1762), the wife of the British ambassador to Turkey. She had lost a brother to smallpox, and she herself was badly scarred from a bout with the disease.

When Lady Montagu arrived in Turkey with her young children, she was very interested in the method used locally to guard against the disease. She observed that the Turkish women held gatherings in September and would invite old women to stop by to offer *variolation* to protect against the disease. The women healers brought scrapings from those who had had the disease. They then selected and cut into four to five veins on each person and used a needle to insert a bit of the smallpox virus and bound the wound with a hollow bit of shell.

Following variolation the person became ill; a few people died from the process, but most had only a light case of the illness. Some remained sick, and during a two- to three-week period the person who was variolated was contagious. This placed the community at a slightly higher risk than an actual outbreak; when people

Lady Mary Wortley Montagu, painting by Charles Jervas *(National Gallery of Ireland and The Yorck Project)*

came down with smallpox under normal circumstances, they were too sick to move about, so they infected very few during the time they were contagious.

Lady Montagu had her children inoculated while in Turkey, and in 1721 when she returned to England she promoted inoculation as a way to keep people safe. For the most part, the British community was disapproving. Then a few months after her return, a smallpox epidemic arose in London, and everyone was frightened. The king was very fearful for his family and was aware of Lady Montagu's experience. When he asked his doctors to variolate the royal family, the royal physicians pronounced the procedure dangerous. The king decided the best course of action was to test the method on others first, and he selected six condemned prisoners to be variolated. (This may have been the first use of humans in experimental trials.) The prisoners suffered mild cases of smallpox, but all fully recovered within two weeks of the exposure and were granted full pardons. The king then tested this on children in an orphanage, and finally suggested his own two daughters be inoculated. This, of course, made it all the rage.

The variolation technique was also used in North America in 1721. It was widely criticized in the beginning, but officials began to note that only six people died out of 244 who were vaccinated, while 844 people died out of 5,980 cases of smallpox in non-vaccinated people. This was clear evidence that vaccination worked.

Fifty years later, George Washington was concerned about vaccinating, knowing that his men would be ill immediately afterward. However, by 1777, a smallpox outbreak occurred, and, when Washington saw the devastation to his army brought about by the illness, he ordered that all men who had not had smallpox should be vaccinated. George Washington was aware that the British reportedly used smallpox to help decimate the Native Americans during the French and Indian War in 1763, and so he also decreed that letters from Boston were to be dipped in vinegar to "cleanse them."

EDWARD JENNER CHAMPIONS A NEW AND SAFER METHOD

By the late 1700s, a British country physician Edward Jenner (1749–1823) championed vaccination using the cowpox virus to prevent the closely related disease of smallpox. Jenner was not the first to come up with this idea. A Dorset farmer—and several other people from similar backgrounds—had induced immunity in their families with cowpox during a smallpox epidemic in 1774. However, Edward Jenner deserves credit for understanding the process and translating it into something that could be used by others. (Jenner studied under the great Scottish surgeon John Hunter, whose advice was always "Don't think, try the experiment," which make his actions very understandable.)

The Cow-Pock—or—the Wonderful Effects of the New Inoculation! Print (color engraving published June 12, 1802, by H. Humphrey, St. James's Street. A British satirist shows Edward Jenner vaccinating frightened young women, and cows emerging from different parts of people's bodies. The cartoon was inspired by the controversy over using "cowpox" to inoculate against smallpox. *(Library of Congress Prints and Photographs Division)*

SMALLPOX TODAY

Despite the creation of a vaccine in 1796, not everyone was vaccinated. Thus, smallpox continued to flare up in various communities upon occasion. By the 1930s, however, the cases of smallpox fell dramatically in the United States and nearly disappeared during World War II. The last smallpox case in the United States occurred in 1949, and routine childhood vaccinations ended in 1972. Five years later, a man in Somalia was the last person in the world to catch smallpox from another human being (he survived). Through an extraordinary international effort, smallpox became the first disease to be completely eradicated.

An unfortunate incident occurred in Birmingham, England, in 1978. A medical photographer Janet Parker contracted the disease while working at the University of Birmingham Medical School, and she died on September 11, 1978. Shortly after, the scientist responsible for smallpox research at the university committed suicide. Based on this

In the late 1790s, Jenner noted that dairymaids of Gloucestershire who had had cowpox (a mild illness) claimed that they would not get smallpox. Jenner took cowpox matter from a milkmaid Sarah Nelmes and vaccinated a young boy James Phipps in both his arms with material from the milkmaid. Phipps suffered fever and some uneasiness but no great illness. Later Jenner inoculated Phipps with the type of smallpox scabs that were being used in variolation, but Phipps—unlike other subjects—did not become sick at all. Jenner tried it again, and still Phipps showed no sign of illness. Jenner tested this on several other people and determined that it worked. In 1798, Jenner wrote *An Account of the Causes and Effects of the Variolae Vaccine.* The medical establishment was slow to accept the theory, but finally, in 1840, the British government banned variolation and provided vaccination free of charge. The accident, all stocks of smallpox were destroyed or stored at one of two World Health Organization (WHO) reference laboratories (one in the United States and one in Russia) where they are to be guarded. In 1986, WHO recommended destruction of the virus, but this did not occur. In 2002, WHO decided against final destruction. Though there is a risk in keeping the virus around, destroying it would prevent the ability to manufacture vaccine in an emergency.

Laboratories in the United States and Russia still maintain tubes of the virus, though many object to this, fearing that it might be stolen and used as a biological weapon. This became a bigger concern after the terrorist attacks of September 11, 2001. In December 2002, George W. Bush received a smallpox vaccination as a beginning push for health care workers to be vaccinated in the event of a biological attack. The campaign fizzled out by the end of 2003 because of problems with the vaccine and a diminishing of public fear of an attack.

idea caught on in other countries, and, in 1803, Spain sent an expedition to the Americas to vaccinate its subjects against the disease. The Spanish reached as far as Santa Fe and vaccinated many settlers and Pueblo Indians.

CONCLUSION

Though no one yet understood the nature of contagion, scientists and physicians—and nonprofessionals such as Lady Montagu—were beginning to make some progress in finding ways to prevent illness. While variolation and vaccination were high-risk experiments when first begun, they eventually proved effective. Elsewhere, simple matters of cleanliness and healthy eating began to improve the health of the people of the day.

5

Learning from Yellow Fever

Yellow fever is an *acute* viral disease that has caused several devastating epidemics. The transmission of the illness is primarily done by mosquitoes, something that was not understood before 1900. Today, the illness continues to occur in Africa and South and Central America and parts of the Caribbean. According to the World Health Organization (WHO) in a 2001 report, yellow fever still causes 200,000 illnesses and 30,000 deaths each year in unvaccinated populations.

Like some other illnesses, childhood cases are generally milder than those suffered by an adult, and those who recover from the illness are immune. Symptoms of yellow fever generally do not emerge until three to six days after exposure, and then there are two phases of the disease. The first is what is termed acute, characterized by fever, muscle pain (largely a backache), headache, shivers, nausea, and vomiting. Then the symptoms disappear, and the person feels better.

In 15 percent of the cases, the illness enters a toxic phase within 24 hours. Fever reappears, and several organs may be affected. The patient appears *jaundiced,* and bleeding can occur from the mouth, nose, eyes, or stomach, and kidney function deteriorates. Half of the patients who enter this phase die within a two-week period.

The others recover without significant organ damage. Treatment for yellow fever is symptomatic and supportive; since it is a viral illness there is no remedy at this point. Because there is no specific treatment for the virus, getting people vaccinated is the key to eradicating the illness.

By the 18th century, yellow fever was one of the most feared diseases in the Americas, because it was so devastating. While mild cases were easily mistaken for influenza or malaria, the more virulent cases caused damage to the heart, kidneys, and liver and often led to death. Statistically, it was not as devastating as tuberculosis or smallpox, but it struck with such ferocity that people were very fearful. In the New World, Philadelphia was one of the areas most affected by yellow fever. Doctors were uncertain of the cause, and this led to more fear.

This chapter will present information on yellow fever in the New World and how society treated it. The misunderstandings of the cause of the illness led to some bizarre solutions, and the esteemed physician Benjamin Rush (1745–1813) was one who had questionable theories about the disease. Until the cause of the disease was determined, it was very difficult to eradicate, but finally—in the late 1800s—the puzzle pieces fell in place so that yellow fever could be better contained.

Philadelphia port *(Philadelphia Water Department)*

YELLOW FEVER IN THE NEW WORLD

Yellow fever existed as early as the fall of Rome, but its reach and devastation expanded with the increase in travel by explorers and conquerors visiting new lands. One of the first major outbreaks during this period occurred in Havana, Cuba, in 1762–63. European and colonial troops arrived in massive numbers to take over the area, in which large numbers of people had no immunity to the disease. The illness struck in epidemic proportions in coastal and island communities in the region, and it recurred frequently over the next 140 years.

How the disease spread was unknown at this time. Scientists today know that the reason the disease occurred primarily in coastal and port cities was because it is spread by mosquitoes, so the more moist the climate, the higher the insect population. The *Aedes aegypti* mosquito, the type of mosquito that carries yellow fever, breeds in freshwater and does well in a populated environment, so some insects may have found freshwater sources on ships traveling in the area, which resulted in the spread of the disease. Yellow fever was also prevalent in Africa, so as explorers arrived in the Americas with slaves, those ships, too, may have been carriers of the disease.

YELLOW FEVER OUTBREAK IN PHILADELPHIA (1793)

Outbreaks occurred in many American port cities, but the one in Philadelphia in 1793 was particularly notable, killing more than 10 percent of the population within a few months. It was the largest yellow fever epidemic in the history of the United States. In the 1790s, Philadelphia had a population of about 55,000 people and was the nation's capital. The summer of 1793 was unusually dry and hot, and because the water levels of streams and wells were low, the shallow, standing water made a perfect breeding ground for insects. By July, the city residents were noting that there were an unusual number of flies and mosquitoes around town.

Philadelphians traded actively with the West Indies, and political turmoil there led to a large amount of refugees from the area arriving in Philadelphia to escape what was happening in their own countries. The Caribbean refugees brought yellow fever with them. While a few mosquitoes may have traveled on shipboard, the arrival of sick people from the Caribbean resulted in infecting the *Aedes aegypti* mosquitoes that resided in Philadelphia. Once they sucked the blood of a person with the illness, these local mosquitoes could then spread the virus to other humans. The disease was first noticed in July, and the numbers grew steadily. Victims initially experienced pains in the head, back, and limbs, accompanied by a high fever. These symptoms would often disappear, leaving a false sense of security. In a percentage of victims, the disease would announce its return with an even more severe fever and turn the victim's skin a ghastly yellow while he vomited black clots of blood. Death soon followed.

Though there had been previous outbreaks of yellow fever, the size of the population and the virulence with which the disease spread caused a high degree of panic. The congressional session was concluding and the nation's leaders were leaving town, but Benjamin Rush, along with other professionals, recommended that as many people as possible leave the city in order to avoid the illness. Those who stayed in Philadelphia responded to the panic by barricading themselves in their houses.

The number of victims taken by the illness grew from 10 a day in August to 100 a day by October, and there were not enough hospitals or physicians to tend to the sufferers. Philadelphia banned newly arriving ships, and the port area was *quarantined,* but other parts of the city continued to operate. The post office was open (but mail delivery was halted); most markets remained open; and in the beginning the churches continued to hold services.

There were a good number of well-respected physicians in Philadelphia at the time, and they were in vehement disagreement as to the correct way to treat the illness. The majority felt that yellow fever should be treated with stimulants including wine and Peruvian bark and the sick people should be bathed in cold water. Benjamin Rush, the city's leading physician and a signer of the

U.S. Declaration of Independence, felt differently. He believed that *calomel* (mercury) purges and copious bleeding with the lancet were the best treatments for the illness.

Mayor Matthew Clarkson wanted to respond to the circumstances, and he worked with the citizenry to establish orphanages, distribute supplies to city residents, and to collect corpses. He also arranged for a temporary fever hospital at an abandoned estate called Bush Hill. Volunteers helped take care of the ill, and the creation of a hospital was made possible by Stephen Girard, a local philanthropist who volunteered to help convert a mansion so that the poor could be taken care of. Bush Hill turned out to be much more successful at curing the ill. Since yellow fever was a virus, some people were able to recover from it. Thus, those who were not put through copious bleedings and purging fared well with palliative care.

A few early frosts in October began to slow the spread of the disease, and by early November the number of cases finally dropped to zero. Scientists today know that the cold weather killed off the mosquitoes, but, at the time, no connection between the disappearance of the insects and a reduction of the disease was made. Business resumed, and Congress resumed its session on schedule in December.

Help from the African-American Community

At the urging of Benjamin Rush, the support of Philadelphia's free black community was enlisted, because it was believed that African black people were immune. Philadelphia's black community put aside their resentment over the way they were usually treated, and they dedicated themselves to working with the sick and dying in all capacities, including as nurses, cart drivers, and grave diggers.

As the weather cooled, the disease subsided, and the deaths stopped. Then accusations began against the black citizens, who had worked so hard to save the sick and dying, of actually being the cause of the spread of the disease. The attack was led by Mathew Carey, whose pamphlet attacked many in the black community. A response to the pamphlet was published by two of the men who had helped organize the black people, Richard Allen and Absalom Jones.

Despite Rush's belief that blacks could not contract the disease, 240 of them died of the fever.

WHAT THEY THOUGHT CAUSED YELLOW FEVER

Today, physicians know that fever is a symptom of some other malfunction in the body, but, at the time, it was believed that if the fever could be eradicated, then all would be well. The spirit of the day also dictated that searching for a "cause" of an illness was a waste of time—the search for a cause of illness was referred to as a "great abyss." They felt that as long as they addressed the fever, it did not matter exactly what caused it.

During the late 1700s, a few scientists were beginning to develop thinking that could have led to an understanding of germs causing illness, but no one ever made the leap. The *Philadelphia Gazette* printed several letters from "Animalcule" who wrote of minute insects causing disease, but no one else picked up on it. The esteemed Rush noted the plethora of insects during bouts of the epidemics, but he felt that the bad air, the miasma, brought them along with the disease.

The general understanding of the cause of almost all illnesses was the theory of miasma, or bad air. (Scientists were beginning to be aware that a dog bite could carry rabies, and a snake bite could kill, but the thought that a very small insect could carry anything lethal to a human being had not yet occurred to them.) Dr. William Cullen, a well-respected professor at Edinburgh who taught many students and influenced their beliefs, taught that fevers were caused by the miasmatic air from marshes acted upon by heat. When Philadelphia experienced an outbreak of dengue fever, Rush felt Cullen's theory was proved. During its occupation of the city during the Revolutionary War, the British army cut down all the trees that had purified the winds blowing up from the marshes, and Rush felt that the dengue fever came from the

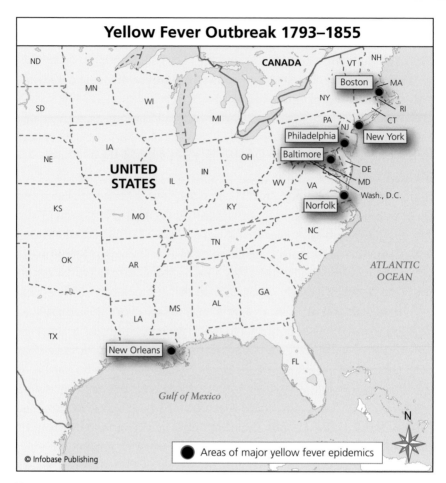

Yellow Fever Outbreak 1793–1855

Areas of major yellow fever epidemics

© Infobase Publishing

Epidemics of yellow fever were common in the Americas. This map shows the cities that were particularly hard hit in North America in the 1790s and the early 1800s.

marshes. Ironically, both dengue and yellow fever are spread by the *Aedes aegypti* mosquito.

Benjamin Rush wrote exhaustively on the appearance of yellow fever. He correctly described the exact stages of the disease, and, with the 1793 epidemic, Rush noted that the hot summer weather had contributed to the bad air. He also felt certain that the specific reason the disease reappeared at this time was because of "morbid vapors" coming from coffee beans that were rotting near the dock: "Mrs. Bradford had spent an afternoon in a house directly opposite to the wharf and dock on which the putrid coffee had emitted its noxious effluvia, a few days before her sickness, and had been much incommoded by it . . ." He then noted that Mrs. Bradford's sister had been exposed when she visited the house to see her sister, and two young boys had spent whole days in a "compting [counting] house" near where the coffee was exposed, and they too had become ill.

Other physicians less powerful than Rush offered that the disease was being brought to Philadelphia by ships from the West Indies where the disease occurred frequently. Though there was disagreement on the cause—and generally speaking they felt the cause was immaterial to the treatment—Rush and others did push for sanitary reform that could have helped improve the situation, although nothing was done differently until several years later.

Theories of Contagion

Once yellow fever occurred in a community, the transmission of the disease was confusing. While scientists and physicians understood that by observing the nature of the illness and the way that it spread one could tell them a lot about a disease, yellow fever was baffling. Physicians in the 18th century defined contagion as something that someone else had that could be transmitted "within a distance of 10 paces." Many people who had no contact with those who were ill with yellow fever came down with the illness; those who cared for the sick did not always get it, and the outbreaks ended with cold weather. Physicians also noted that people who fled did not carry the disease with them.

Those who believed it was contagious undertook to inoculate themselves with blood, vomit, or saliva of yellow fever patients. (They knew that vaccinating against smallpox had been successful; see chapter 4.) Those who believed it was contagious felt the sick were to be feared, and, as a result, the sick often died of neglect because no one would go near them to offer help.

In some circles, the arguments became political. Federalist physicians believed yellow fever was contagious and had been imported from Haiti by French refugees. They felt that blaming miasma—local conditions—was unpatriotic. They advocated quarantine, limiting trade, and felt that healing would take place if they gave patients quinine and wine. Jeffersonian (Republican) doctors did not feel yellow fever was contagious. They felt yellow fever arose from unsanitary local conditions, and they fought against quarantine regulations and restrictions on trade.

Noah Webster (1758–1843), who is best known as an American lexicographer, sent questionnaires about the nature of the illness to physicians in Philadelphia, New York, Baltimore, Norfolk, and New Haven. In 1796, he published results and his conclusions—*A Collection of Papers on the Subject of Bilious Fever, prevalent in the United States for a Few Years Past.*

In Europe, they used other cures for yellow fever, but they were not necessarily milder. The well-regarded French clinician P. C. A. Louis (1787–1872) detailed what was done to a patient under French care: Day one, the patient was given a large dose of castor oil, an enema, and several doses of calomel while leeches were applied to his temples. On the second day, the patient received calomel and then was bled by leech and lancet. On the third day, he was given several enemas and 25 drops of laudanum; he died before the end of the third day.

Some 18th-century scientists opted to further investigate the disease and explored the chemistry of the vapors from the disease, but they really did not know what they were looking for. Some doctors decided that tasting the black vomit would provide helpful information, and they did so—and survived. Though they did not learn anything about the chemical composition of the disease, this process did prove that vomit did not transmit the fever.

THE LEGACY OF BENJAMIN RUSH

Benjamin Rush presents an interesting puzzle for historians. He figures prominently in the discussion of yellow fever because he

RUSH'S CONTRIBUTIONS
TO MENTAL HEALTH

There are those who argue that Rush's approach to clinical medicine was correct—that he observed what was happening and acted accordingly, and this placed science in the United States on an improved path. However, all agree that Rush did not do well at divining how disease should be treated. Bloodletting was not the correct answer. The clinical contributions of Benjamin Rush will continue to be debated, but there is one area where Rush had a more positive impact, and that was in the field of mental health. He is sometimes referred to as the father of American psychiatry, and his image appears on the American Psychological Association (APA) seal. He published the first textbook on the subject in the United States, *Medical Inquiries and Observations upon the Diseases of the Mind* (1812).

Rush was notable for his time in believing that mental illnesses could be cured. Mental patients at that time were often locked up or beaten, and Rush helped society move away from this horrific type of treatment as he listened to the patients. Unfortunately, as with his faith in bloodletting, Rush developed some questionable practices with these patients as well. He believed mental illness could be forced out of a person, and so one of his treatments for psychiatric illness involved tying a patient to a board and spinning it rapidly so that all the blood would go to the person's head. He also devised chairs that were suspended from the ceiling, and attendants were assigned to supervise and swing and spin the mentally ill patient for hours. A tranquilizer chair he created in 1811 locked a person into a chair in a place where all light could be cut off, much like the sensory deprivation tanks of today.

On a more positive note, Rush was the first to understand the nature of addiction and further carried the idea that abstinence was the only cure.

Benjamin Rush *(Dibner Library of the History of Science and Technology)*

wrote a great deal about the disease, studied it carefully, and bravely stayed in Philadelphia throughout times when the disease ran rampant because he was dedicated to caring for those who were sick. He wrote copiously about illness, shared his research freely, and encouraged research by others, and he was the first to identify the fever that was spreading as yellow fever. He convinced others that "something had to be done" for the ill, since their deaths were not an order of the "hand of God." He also paved the way toward better treatment of those who suffered mental illness (see the previous sidebar, "Rush's Contributions to Mental Health").

In addition to all these attributes, Rush was a true American hero. He was an important American political leader who had been vital to the break with Britain for independence. He was a member of the Continental Congress and a signer of the Declaration of Independence and held many different government positions to help keep the young country on solid footing. He was also appointed treasurer of the United States.

Yet his practice of "heroic medicine," bleeding and purging the ill and dosing them with calomel, undoubtedly caused the deaths of many who might have gotten better if they had simply been permitted to ride through the dreadful illness.

HEROIC MEDICINE

In treating one of his early cases of "bilious fever," he was called to see Polly, wife of Thomas Bradford. Rush switched from mild herbal purges to a harsher chemical calomel or mercurous chloride. He then had a bleeder take out 10 ounces (0.30 l) of her blood to cure the inflammation, and Polly recovered. Rush went on to continue this type of treatment with others. He believed that patients should eat little and then be subjected to vigorous purges with calomel and *jalap,* and bleeding until the patient fainted. He sometimes removed a quart (0.95 l) of blood at a time and repeated this type of bleeding two or three times within a two-day period if the person did not get better. He believed that draining up to four-fifths of a person's total blood would be all right.

Rush was not alone in his belief in this type of heroic medicine. A British contemporary Dr. William Buchan wrote in his book *Domestic Medicine*:

> In this and all other fevers, attended with a hard, full, quick pulse, bleeding is of the greatest importance. This operation ought always to be performed as soon as the symptoms of an inflammatory fever appear. The quantity of blood to be taken away, however, must be in proportion to the strength of the patient and the violence of the disease. If after the first bleeding the fever should rise, and the pulse become more frequent and hard, there will be a necessity for repeating it a second, and perhaps a third, or even a fourth time, which may be done at the distance of twelve, eighteen or twenty-four hours from each other, as the symptoms require. If the pulse continues soft, and the patient is tolerably easy after the first bleeding, it ought not be repeated.

Rush became an ideal target for a British journalist in the United States. William Cobbett (1763–1835) established *Porcupine's Gazette* and attacked Rush in print for his "heroic treatments." Cobbett undertook studies of the Philadelphia yellow fever epidemic and accused Rush of an unnatural passion for taking human blood. Cobbett wrote that many of Rush's patients actually bled to death. According to Cobbett, Rush's method was "one of those great discoveries which have contributed to the depopulation of the earth."

(continues on page 86)

WALTER REED, M.D. (1851–1902):
An Enlightened Approach

Major Walter Reed was a U.S. Army physician who made many contributions to medicine; one of the most significant being that he helped devise a way to prevent yellow fever. This was a major step forward in biomedicine, and his work eventually made the construction of the Panama Canal possible.

Reed completed his medical training for his first degree in 1869 at the age of 17 and went on to enroll at New York University's Bellevue Hospital Medical College to get a second medical degree. He interned at several different hospitals and worked for the New York Board of Health and then took an assignment with the U.S. Army Medical Corps, where he was primarily assigned to posts in the western parts of the United States. Later, he completed advanced work in pathology and bacteriology.

During the Spanish American War, the U.S. military became highly aware of the problems of yellow fever; thousands of American soldiers who fought in this war became ill or died from it. In 1899, Reed was sent to Cuba to study army encampments and the spread of disease, and, the following May, he was specifically assigned to examine tropical illnesses, including yellow fever. In Cuba, Reed's commission conducted a series of experiments on human volunteers, some of whom were medical personnel. To Reed's great sadness, Jesse William Lazear, Reed's own assistant, died from the testing the group was conducting.

Reed's commission eventually identified that mosquitoes led to the spread of yellow fever. While Reed is usually credited with this discovery, Reed himself credited the Cuban physician Dr. Carlos Juan Finlay (1833–1915). By the end of

Reed's time in Cuba, the U.S. Army Yellow Fever Commission confirmed Finlay's theory, and they also demonstrated that it could not be transmitted by clothing or blankets, something that was beginning to worry army officials.

Walter Reed *(National Library of Medicine)*

Reed's conclusions led other scientists to devise ways to reduce the mosquito population so that work could continue in the Panama Canal Zone, something that had been impossible up until this time. From 1881–89, the French who were working to construct the pathway through North and South America suffered a very high death toll from disease; as many as 22,000 workers were estimated to have died during that period. As the United States began to conquer the mosquito, they saw a drop in both yellow fever and malaria. Finally, significant work on the Canal was possible.

Reed returned from Cuba in 1901 and died in 1902. In 1909, Congress ordered construction of Walter Reed General Hospital, which is now known as the Walter Reed Army Medical Center. It is the army's primary medical center on the east coast.

(continued from page 83)

Rush sued for libel, and the popularity of Benjamin Rush, as an American hero, combined with no science to prove that he was wrong, led the jury to find in favor of Benjamin Rush. Cobbett was fined $5,000, a very considerable sum for that time.

CONCLUSION

Yellow fever was a baffling disease, and, since 18th-century physicians had no understanding of how it spread, they were particularly inept at devising treatments. While Benjamin Rush will long be a name that is associated with important events concerning American history, he lived at a time when the links between cause and disease were not well understood. It was not until the turn of the 20th century that answers were assembled in such a way that diseases could be better contained. As a result, world travel gained a shortcut as ships could finally make their way across the isthmus of Panama.

6

Early American Medical Care

Medicine in early America was much like it had been in Europe just after the Middle Ages. The country was made up of agricultural communities, and, although a few physicians arrived in America after studying in Edinburgh or London, the number of trained medical professionals was very small. In some towns, pharmacists were available to prescribe and sell remedies. Surgical issues were taken care of by barber-surgeons, some of whom traveled from community to community to pull teeth and set bones.

Most of the family health care was overseen by the women in the family. Although they rarely received any type of formal education, these women filled the roles of doctor, nurse, and pharmacist for family, servants, and neighbors. Childbirth generally required the help of others who had been through it, and, according to the historian Susan Norwood, women of colonial times were often pregnant and childbirth was a frightening time—many mothers and babies did not survive.

Home remedy books offered guidance on medical issues, and those who were literate might turn to *The Housekeeper's Pocket-Book* by Sarah Harrison or *Every Man His Own Doctor* by John Tennant or *Primitive Remedies* by John Wesley. Wesley's remedy involved a healthy diet, fresh air, plenty of exercise, and simple medicines. Medicinal recipes were also contained in many cookbooks of that time. Those who could not read obtained information from family members.

Medicinal herbs were popular in the colonies. Many had heard of Dr. Nicolas Culpeper of England whose thinking on herbal cures was still well regarded, but the colonists also noted how the Native Americans used plants for specific things. The young women of the colonies learned from their mothers how to grow the herbs they needed. A few specialty items needed to be obtained from apothecaries. One of the more popular remedies available from apothecaries was calomel, a form of mercury. Today, mercury is known to be a neurotoxin, but it was years and years before anyone knew it was dangerous to use. Europeans who visited America often went home and discussed the terrible state of American health. Many experts feel this was due to the amount of mercury that was consumed in this country.

At the beginning of the 1700s, there was little understanding of the importance of personal hygiene and its role in keeping people healthy. By the middle of the 19th century, the middle and upper classes began to think that washing one's hands and face in the morning was a good idea, and women began to change their skirts a little more often. (Up until this time, women tended to wear the

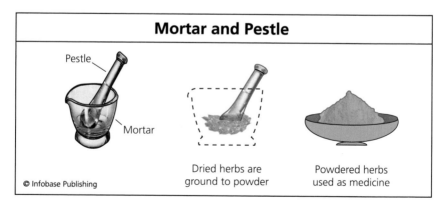

Mortar and Pestle

Pestle

Mortar

Dried herbs are
ground to powder

Powdered herbs
used as medicine

© Infobase Publishing

These were the tools apothecaries and healers used to crush, grind, and mix substances into medicinal cures.

same thing without washing their clothing, so the skirt dragged around in the dirt, and it rarely was washed.) Despite these changes, few people brushed their teeth, which contributed to a high rate of tooth loss and gum problems, and the custom of washing babies' diapers did not occur until just before the Civil War.

This chapter will examine the state of early American medical care, from the theory behind "heroic medicine" to the qualifications and training of the physicians who came to America right through and including the creation, popularity, and use of patent medicines.

EARLY AMERICAN PHYSICIANS

In 1775, there were approximately 3,500 practicing physicians in the colonies. Some had been trained at the Pennsylvania Hospital in Philadelphia, the first medical college to open in America in 1768. This institution was followed by King's College (now Columbia University), which opened two years later in New York. Because these colleges accepted only a handful of doctors for training, most American doctors were trained through apprenticeships, receiving seven years of training before they were officially considered physicians. While these doctors were highly trained by the

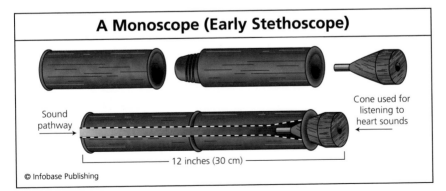

A Monoscope (Early Stethoscope)

Sound pathway

Cone used for listening to heart sounds

12 inches (30 cm)

© Infobase Publishing

The monoscope was an early form of a stethoscope that was used to listen to sounds from the chest of someone who was ill.

standards of their time, their services were not available to all of the general population. Many people lived too far away from any doctors to use their services, and other people did not have access to doctors because of social customs or beliefs. For these reasons, other types of healers often assumed the role of caring for the injured or sick.

Around the time of the Revolutionary War, a movement of "Thomsonians" cropped up. The group was named after the physician Samuel Thomson (1769–1843), who gave up an orthodox practice based on bloodletting to develop a much simpler theory based on steam baths and the Indian herb, lobelia. If a person consumed enough, the result was a purging of their system.

As the number of physicians began to increase, it became necessary to limit who could practice medicine. In 1806, the nation's first licensing law, the Medical Practices Act, was passed in New York State. This did not place specific limits on who could be a healer, but it restricted who had legal rights if there was a problem. The Medical Practices Act permitted only licensed physicians to recover their fees in court, and, if caught, unlicensed practitioners were fined $25 for practicing without a license. The *New York Journal of Medicine* praised this movement and noted that this would suppress empiricism and encourage the growth of regular practitioners. However, this did not happen quickly. America represented the land of freedom, and lay practitioners campaigned against the

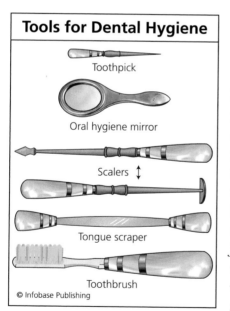

Tools for Dental Hygiene

Toothpick

Oral hygiene mirror

Scalers ↕

Tongue scraper

Toothbrush

© Infobase Publishing

The importance of oral hygiene was beginning to be understood, but not everyone practiced it. This type of set would have belonged to a dentist (or barber-surgeon) who treated an upper-class clientele.

new law. In 1807, the Medical Practices Act was modified so significantly that it was virtually repealed, and eventually the law was abolished in 1844. This same pattern of lawmaking and repeal followed in Connecticut and Massachusetts as well as a good number of other states. Finally in 1847, the American Medical Association (AMA) was created and provided a national platform to promote the interests of the profession. Although the AMA became an important organization, by the end of the 19th century, state licensing agencies finally gained strength and could regulate what was happening in their states.

The field of dentistry officially established itself before the Civil War. The first American dental journal was published in 1839, and in 1840 the Baltimore College of Dental Surgery and the American Society of Dental Surgeons were founded.

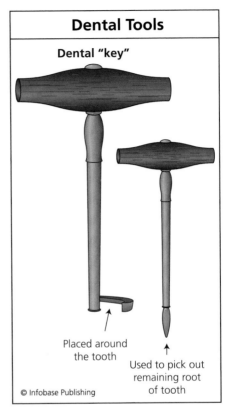

Dental Tools

Dental "key"

Placed around the tooth

Used to pick out remaining root of tooth

© Infobase Publishing

These were very crude tools that a barber-surgeon or tooth puller would have used to extract a tooth during the early 19th century.

THE AGE OF HEROIC MEDICINE

During the age of heroic medicine (1780–1850), educated professional physicians aggressively practiced bloodletting (venesection), intestinal purging, vomiting, profuse sweating, and *blistering*. Massive doses of drugs were used in the belief that "more was better." Physicians originally treated diseases like syphilis with salves made from mercury. These medical treatments were well-intentioned and often well accepted by the medical community, but were actually harmful to the patient. (The term *heroic medicine* was used pejoratively about 100 years later by Oliver Wendell Holmes. He realized that it was the patient who needed to be heroic, not the physician.)

There was no real ability to run clinical trials at this time, and when a scientist did try to prove something did not work it was never well accepted. One who tried to prove that bloodletting was pointless was the French physician Pierre-Charles-Alexandre Louis (1787–1872), who used a numerical system to collect information from hospital patients to evaluate therapeutic methods. His studies of venesection showed that it was ineffective if not harmful, but they had little impact. Critics blamed those who were running the studies and doing the diagnosis. The spirit of the day led many to argue that the reason venesection was not working was because it was not being done aggressively enough.

By 1825, homeopathy finally caught on in the United States. This medical theory that honored natural healing with an emphasis on nutrition and exercise and recommended the use of minute amounts of medicine was a response to the heroic movement.

Medical Tools of the Day

The average 18th-century physician had little in the way of either equipment or understanding to aid him in distinguishing one specific disease from another. The concept of a standard body temperature had only been suggested, the body's heat-regulating mechanism was not understood, and Fahrenheit's recently developed mercury thermometer was not commonly used by physicians.

The stethoscope was not invented until 1814 by René Laënnec, who practiced medicine in Paris. In 1819, he wrote about the many and curious sounds of the heart and lungs that can be heard using his instrument. About the same time, Leopold Auenbrugger, a Viennese physician, discovered the use of "percussion"—the use of the fingers to tap on the chest—to investigate diseases of the chest. He was the son of an innkeeper and had learned this method to judge the level of wine in his father's casks.

A "pulse watch" had been developed in 1707, but it also was largely ignored by physicians, who preferred describing the pulse to counting it. The use of percussion to aid in diagnosis, however, was beginning to become more widely understood because of the work of Leopold Auenbrugger.

THE MEDICAL SIGNIFICANCE OF GEORGE WASHINGTON'S DEATH

George Washington's death is very instructive concerning how medical issues were perceived and treated at the end of the 18th century. No man was more revered, so the treatment he received would have been the best available at the time. While there is some disagreement as to the cause of his illness, medical professionals are in agreement that the actions taken would not have led to any sort of cure.

On December 12, 1799, Washington spent several hours inspecting his farms on horseback, in snow and hail and freezing rain. The following day he noticed his voice was hoarse and his throat sore, but he went out again to mark trees that needed to be cut. During the early morning hours of Saturday, December 14, Washington awoke and felt quite poorly and was having difficulty breathing. He would not let Martha summon help until sunrise. The maid arrived to light the fire and reported that he was in serious respiratory distress. His aide, Colonel Lear, was summoned, and Lear sent for the estate overseer who prepared a medicinal mixture of molasses, vinegar, and butter, but Washington was unable to swallow it, so they decided that bloodletting was a remedy that would be helpful. The overseer opened one vein, though Martha, who did not believe in venesection, objected. His feet were then bathed in warm water.

By midmorning, Washington's physician, Dr. James Craik, arrived, and he also sent for two other prominent physicians, Dr. Gustavus Richard Brown and Dr. Elisha Cullen Dick. Dr. Craik created a blister of *cantharides* (dried beetles) to place on his throat and took out 20 ounces (0.59 l) of blood from two separate bloodlettings.

Dr. Dick arrived at 3:00 P.M. and took out 32 ounces (0.9 l) more of blood from Washington's forearm. Dr. Brown arrived, felt

Washington's pulse, and the three physicians decided that *tartar* and calomel should be administered rectally.

By late afternoon, Washington began to realize that nothing was working, and he calmly began giving Colonel Lear instructions as to what to do after his death. At 8:00 P.M., the physicians applied blisters and *poultices* of wheat bran to his legs. Dr. Dick wanted to perforate Washington's trachea to improve his breathing, but it was a very new procedure and the other physicians discouraged it.

A few minutes before 10:00 P.M., Washington asked the time and then, according to an eyewitness account written by Martha Washington's grandson (by her previous marriage to Daniel Custis), George Washington Parke Custis, who was with the family at the time of death, "He spoke no more—the head death was upon him, and he was conscious that 'his hour was come.' With surprising self-possession he prepared to die. Composing his form at length, and folding his arms on his bosom, without a sigh, without a groan, the Father of his Country died."

Because of Washington's importance to the country, there were many contemporary reports on his death, including a detailed description of the treatment that was written by Drs. Craik and Dick and published in the *Times of Alexandria* on December 19, 1799. This has permitted modern scholars and medical professionals to examine the details of his illness as well as to consider his treatment. While some reports say that Washington died of acute laryngitis that became pneumonia, others feel his illness was *quinsy,* a peritonsillar abscess, which is a recognized complication of tonsillitis and consists of a collection of pus beside the tonsil. Others wrote of "acute inflammatory *edema* of the larynx." These diagnoses might explain the early difficulty with both breathing and swallowing. In addition, the case may have been complicated by infection from streptococcus or some other type of bacteria that would have been difficult to wipe out without antibiotics. In 1997, an otolaryngologic publication published a piece that noted that it was acute bacterial epiglottitis, which can obstruct the airflow and cause a suffocating death.

THE DISCOVERY OF IODINE

Iodine was discovered in the early 19th century by a fellow named Bernard Courtois whose family business made saltpeter, which was important in the manufacture of gunpowder. Courtois was working to isolate sodium carbonate from seaweed so that it could be used to create saltpeter, and when he added too much sulfuric acid he created iodine. Intent on what he was making, he did not stop to explore what had happened, but he gave samples to several of his friends, including Louis Gay-Lussac (1778–1850), a well-respected chemist, and another chemist, Humphry Davy (1778–1829). Gay-Lussac announced that the substance was either a new element or a compound of oxygen, and, around the same time, Davy sent a letter to the Royal Society of London stating that he had found a new element. A large argument erupted between Davy and Gay-Lussac over who identified iodine first, but both scientists eventually acknowledged Courtois was the first to isolate it.

Sir Humphry Davy made many important contributions to chemistry during his lifetime, including discovering the elemental nature of chlorine and iodine. *(National Library of Medicine)*

Iodine is important in two ways to the study of medicine. It is an essential trace element needed by living organisms; animals and humans need iodine for their thyroid hormones to function properly. In addition, elemental iodine was discovered to be helpful in sanitizing things—an advance that was going to become more and more important as scientists learned how important cleanliness in wound care and surgical procedures would be.

Regardless of how serious Washington's illness was, all agree that the treatment methods chosen—typical for the day—were ineffective and possibly harmful. Some go as far as to suggest that it was the treatment that killed him.

Dr. Vibul V. Vadakan, a Los Angeles pediatric hemato-oncologist, has done detailed studies of Washington's treatment, and his study reveals that 82 ounces (2.4 l) of Washington's blood was removed during a 13-hour period. Taking into consideration Washington's height and weight, Vadakan notes that the blood removed would have been more than half his blood volume during a short time. This would have led to preterminal anemia, hypovolemia, and *hypotension.* His calmness at death leads Vadakan to note that he was likely in shock from blood loss.

MEDICINES BECOME PRODUCTS

Patent medicines originated in England during the 1600s as proprietary medicines manufactured under grants, or "patents of royal favor" to those who provided medicine to the royal family. These medical tonics were sometimes referred to by their Latin name, *nostrum remedium* (our remedy), and this was then condensed down to *nostrums.* Those who received a royal patent were given the exclusive right to make a particular remedy.

To apply for a patent, the remedy needed to be unique, but there were no standards concerning whether it was effective or safe. One of the first patent medicines was Robert Turlington's "Balsam of Life." (See the following sidebar, "The Story behind the Balsam of Life" on page 98) Some of the other remedies sold to the public included Anderson's Pills, Lockyear's Pills, Dr. Bateman's Pectoral Drops, Daffy's Elixir Salutis for "colic and griping," and Dr. John Hooper's Female Pills. As the British moved to America, they brought the concept of patent medicines, with them, and these cure-alls were sold by postmasters, goldsmiths, grocers, and tailors.

After the American Revolution, royal endorsement became valueless, and because getting a patent required listing the ingredients, many owners never applied for a real patent. Instead, they registered distinctive trade names so that they could claim exclusive ownership over whatever it was they were selling.

Eventually—by the 19th century—all medical preparations came to be known as patent medicines. All featured "colorful names and colorful claims." Many promised magical healing cures while others referred to some type of medical discovery. The experts usually did not exist, and the names were selected to conjure an exotic locale or to promise an incredible cure. Those who sold the medicines and masqueraded as the experts usually claimed to be Indian or Oriental as these people were thought to have mystical knowledge. A few claimed Quaker heritage because the Quakers were known to be scrupulously honest. Sometimes, snake oil salesman was used to describe these fellows. While the term now has a connotation that is not complimentary, its original use

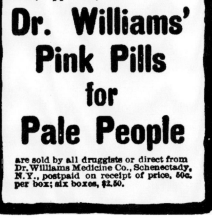

Miraculous Cure

Richard D. Creech, of 1062 Second St., Appleton, Wis., says:

"Our son Willard was absolutely helpless. His lower limbs were paralyzed, and when we used electricity he could not feel it below his hips. Finally my mother, who lives in Canada, wrote advising the use of Dr. Williams' Pink Pills for Pale People and I bought some. This was when our boy had been on the stretcher for an entire year and helpless for nine months. In six weeks after taking the pills we noted signs of vitality in his legs, and in four months he was able to go to school. It was nothing else in the world that saved the boy than Dr. Williams' Pink Pills for Pale People.—*From the Crescent, Appleton, Wis.*

Dr. Williams' Pink Pills for Pale People

are sold by all druggists or direct from Dr. Williams Medicine Co., Schenectady, N.Y., postpaid on receipt of price, 50c. per box; six boxes, $2.50.

Miraculous Cure, an advertisement for Dr. Williams' Pink Pills for Pale People, an old patent medicine

THE STORY BEHIND THE BALSAM OF LIFE

The British merchant Robert Turlington invented a balsam for which he obtained a patent from King George II in 1744. According to records of the British Patent Office, Turlington's unique patent medicine consisted of 27 ingredients, and the patent gave him the right to "make, use, exercise, and vend" the said specific balsam. Patents also provided patent holders with proprietary rights over the product, which could involve pursuing imitators or anyone who claimed to be selling his product but were actually selling a substitute.

Turlington's Balsam of Life quickly gained fame and was used in households everywhere; it was particularly popular in the American colonies. Though there were stiff penalties for violating the terms of a royal patent, the increased popularity of Turlington's Balsam encouraged others to pirate his idea. Some copied the bottle and distributed their own brew, others bought up used Turlington bottles and filled them with their own concoctions.

Turlington had a lot at stake, and so he actively pursued his right to maintain control of the product. Originally the Balsam was packaged in a relatively simple medicine bottle, but Turlington realized a unique bottle design would make the Balsam more difficult for imitators to copy. In 1754, he selected a blown molded bottle that was slightly pear-shaped; the name Balsam of Life appeared in embossed lettering. The new design was much more difficult to copy.

Turlington's successors were able to maintain the remedy's popularity, and the product sold for many years. Much later, however, the Balsam of Life was affected by the U.S. Pure Food and Drug Act passed by Congress in 1906, and the company stopped selling the Balsam under its longtime name. They continued to market the medicine, but gave it another name—Compound Tincture of Benzoin (a.k.a. tincture benzoin compound).

stems from the most common claim of patent medicines—that they contained oil extracted from a snake.

Patent Medicine Ingredients

Patent medicines rarely contained what they said they did. Most did not even contain anything that was even close to what was claimed. A medicine called Vital Sparks was rock candy rolled in powdered aloe. Tiger Fat proclaimed that it was made from backbones of Royal Bengal tigers but was actually made of Vaseline, camphor, menthol, eucalyptus oil, turpentine, wintergreen oil, and paraffin. Dr. Kilmer's Swamp Root was made with unspecified "roots found in swamps" and was touted as helpful to the kidneys. Some medicines for "female complaints" actually contained pennyroyal, tansy, and savin—ingredients that could cause an abortion. One called Liver Pads, to cure problems of the liver, consisted of fabric with a dot of glue on it. The glue itself was laced with cayenne pepper and, when it was pressed against the stomach, the body's heat melted the glue, and, as the pepper permeated the skin, it created a burning sensation that people interpreted as curative.

Kickapoo Indian Sagwa, Blood, Liver, and Stomach Regulator, an old patent medicine advertisement

Others were far from harmless. Jakeleg, an extract from Jamaican ginger, was a patent medicine that was later doctored with a chemical that affected the nerves; some who drank it lost use of their hands and feet.

Most medicines contained a high alcohol base; some contained cocaine, opium, or morphine, and while these ingredients may have

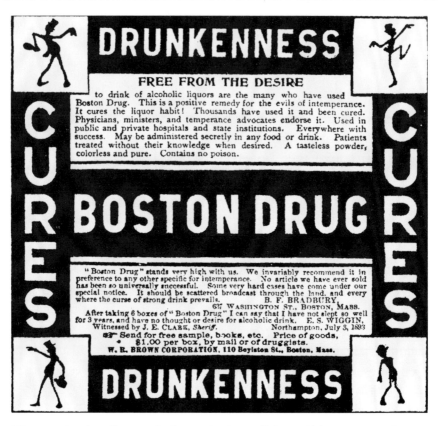

Like most advertisements for patent medicines, this one promises a miraculous cure for drunkenness if Boston Drug is taken.

temporarily numbed a patient's pain, there was no control or supervision over their creation and no authoritative guidance on their dosage, meaning that they could be dangerous. Many people became addicted to those that had opiates or alcohol or cocaine. One, Kopp's Baby Friend, was sweetened water and morphine, and while it was guaranteed to "calm your baby," there are no statistics on how many infant deaths it caused.

Prohibition was a wonderful boost for the patent medicine business. Lydia E. Pinkham's Vegetable Compound was approximately 15–20 percent alcohol, and Dr. Hostetter's Stomach Bitters contained 44.3 percent alcohol, which made it more potent than

80-proof whiskey. Prohibition forced many alcoholic drinks to be for "medicinal use" only.

How They Were Sold

In earlier times, people who practiced healing or those who ran apothecary shops were the primary purveyors of patent medicines. While there have always been quacks and charlatans, the era of patent medicines changed all that—these remedies were the first major product sold via advertising, and the guise of a medical professional was dropped in favor of the professional huckster—the medicine man. The experts who "invented" the cure-alls frequently did not even exist. In some cases, however, real people may have felt that they had a real cure.

For the most part, it was all about the sale, and patent medicines led to the growth of the newspaper industry. Benjamin Franklin's *Pennsylvania Gazette* ran an ad as early as 1731 for the "Widow Read's Ointment for the Itch," which had been created by Franklin's mother-in-law. Later, a fellow named Guy Gannett eventually had a chain of newspapers (not related to the current Gannett newspapers) that started in Maine with a publication called *Comfort* to promote Oxien, made from the African baobab tree.

Free almanacs also became a way of promoting patent medicines. One of the first to use this method was William Swaim who ran a six-page ad for his *panacea*; in the 1832 Farmer's and Mechanic's Almanac. The ad was a treatise on the benefits of Swaim's Panacea, which carried the subtitle: "Being a Recent Discovery for the Cure of Scrofula or King's Evil (primary tuberculosis), Mercurial and Liver Disease, Deep-Seated Syphilis, Rheumatism, and all disorders arising from a Contaminated and Impure State of the Blood." The ad featured cases illustrating the success of the remedy that largely consisted of sarsaparilla, oil of wintergreen, and mercury. Swaim's use of the almanac to promote his product was a first and became widely copied. Free almanacs began to be created to tout the benefits of whatever drug was being backed by the almanac. Dr. David Jayne launched a Medical Almanac and Guide to Health in the 1840s to push such medicines as Jayne's Sanative Pills, Jayne's Vermifuge, and Jayne's Alternative. Dr. Jayne's Alternatives promised to cure at least 25 different illnesses, ranging from cancer to skin problems.

Later on, medicine shows became a popular way to sell the medicines. Medicine men traveled from town to town to present circus and vaudeville-style entertainment. Four to five times during each hour of entertainment, the medicine man interrupted the performers to present a sales pitch. After the show the entire group moved on to the next town. With this kind of operation, they did not have to worry about whether the medicine worked because they moved on to the next community before customers had time to become outraged.

By 1881, the medicine show sales techniques were perfected, and the most successful shows were produced by John E. "Doc" Healy and "Texas Charlie" Bigelow for the Kickapoo Indian Medicine Company. The entertainment was halted four to five times each hour so that the pitchmen could talk about the medicine. A *shill* would be in the audience, buy the product, and then declare himself cured. Salesmen who circulated through the crowd would only carry a few bottles with them so that they could make a big show of "selling out" and having to run backstage for more product. It was a highly effective sales technique for products with little curative merit, though customers may have been sedated from them; some became addicted. In 1859, $3.5 million of patent medicines were sold; by 1904, the earnings rose to $74.5 million.

BRINGING AN END TO THE TRADE

By end of the 19th century, Americans were finally becoming skeptical. Some physicians had always spoken out against these medicines, feeling that they did not work and kept people from seeking legitimate help. The temperance movement also campaigned against patent medicines because of the amount of alcohol in them, and, over time, a public sentiment began to favor laws that disclosed the ingredients. Manufacturers were concerned at the very real thought of lost revenue, and in 1881 they formed the

Proprietary Association, a trade association of medical produc-
ers. The Association received support from the press, which had
grown dependent on remedy advertising. The newspaper and mag-
azine business benefited greatly from the advertising income, and
the patent medicine makers had a clause in their advertising contracts that noted that if the medicines ever became regulated, the
advertising contract would be considered void. This was enough
to quiet the one group that might have been the logical people to
bring attention to what was happening.

Finally in 1892, *Ladies' Home Journal* made a bold move and
quit accepting ads for patent medicines. The muckraker and
reporter Samuel Hopkins Adams, a writer for *Collier's Weekly* in
the early 1900s, wrote "The Great American Fraud," which pub-
licized the problems and the deaths that had occurred from patent medicines. When the article appeared in 1905, this spurred a
campaign against patent medicines and their false claims, noting
the high quantity of alcohol and opiates were "undiluted fraud."

With strong support from President Theodore Roosevelt, a
Pure Food and Drug Act was passed in Congress in 1906, creating the Food and Drug Administration (FDA). The original
act paved the way for public health action against unlabeled or
unsafe ingredients, misleading ads, quackery, and other rackets.
It did not ban the ingredients or the sale of the medicines; it only
stated that they had to be accurately labeled. The Pure Food and
Drug Act of 1906 imposed regulations on the labeling of products
containing alcohol, morphine, opium, cocaine, heroin, chloroform, cannabis, and similar ingredients. It required that products
containing any of those substances be labeled with the substance
and quantity on the label. Use of the word *cure* for most medicines was nominally prohibited. Soon, cure was replaced by remedy and other terms.

The Pure Food and Drug Act was strengthened with the passage of the Sherley amendment in 1912. According to the FDA
Web site, Congress enacted the Sherley amendment to prohibit
the labeling of medicines with false therapeutic claims intended
to defraud the purchaser, a standard difficult to prove. The use of

the word *cure* was largely curtailed, and this is for all intents and purposes the end date for patent medicine bottles for human use that are embossed (or labeled) with cure. However, enforcement was still not specified, and some use of the term most likely did occur after 1912–13, although not likely embossed on bottles after this point. One of the first patent medicines prosecuted in 1913 was William Radam's Microbe Killer, whose bottles claimed boldly to "Cure All Diseases."

A number of patent medicines were still available as late as the 1950s, sold under slightly different names, and today a few of these medicines have morphed into something that is still on shelves: Smith Brothers Cough Drops, Geritol, Absorbine, Bromo-Seltzer, Carter's Little Pills, Luden's, Phillips Milk of Magnesia, Lydia E. Pinkham's Vegetable Compound, Vicks VapoRub. Among the products that are still available in soft drink form but began life as a patent medicine are Hires Root Beer, Coca-Cola (the original contained cocaine), 7-Up, Dr. Pepper, and tonic water (which still contains quinine).

Louis Pasteur made many contributions to science and medicine. One of his early discoveries was a process now known as pasteurization. Engraving by Heliogre Dujardin *(Dibner Library of the History of Science and Technology)*

Ironically, Louis Pasteur's scientific germ theory of disease was introduced to America by patent medicine sellers. One of the main ones was William Radam, a Prussian émigré who lived in Texas. He was interested in Pasteur's discovery of the microbe, and Radam developed a medication to fight these entities. He patented the Microbe Killer in

1886. While it was extremely popular at the time, a later chemical analysis revealed that it was 99 percent water and therefore of no clinical value.

CONCLUSION

The state of early American medicine was quite poor. Physicians lacked formal education and had little clinical training and their misunderstandings about the nature of disease led them to assume that more was better—a treatment that was helpful only if the patient was lucky. Patent medicines offered an opportunity to improve the state of medicine, but because the profit motive quickly outweighed any true interest in healing, the concoctions were thrown together in such a way that people either invested in mixtures that were little more than water or they purchased something that calmed them—and possibly addicted them to the medicine.

Patients who were tended to at home by loved ones and encouraged to rest and eat well were sometimes fortunate. However, their fate depended largely on the virulence of whatever it was that had laid them low. Survival was a matter of good fortune.

7

Early Thoughts on Digestion and Respiration

Today, physicians know that people need to be able to breathe clean air, get exercise, and eat well-balanced meals, but during the 18th and early 19th centuries, there were many misunderstandings about what contributed to good health, and these misunderstandings were largely rooted in ignorance of the bodily processes.

Though studies of anatomy identified the work done by the lungs, no one really understood how the body used air. However, they had their theories about good air and bad air. During the day, a fresh breeze blowing through the house was thought to be a tonic; at night, it was unhealthy, and people covered their windows and locked their doors at night to keep out the bad air.

Proper nutrition was also a big mystery. During the Revolutionary War, Washington felt that the reason his men were sick so much of the time was that they were eating too much meat and not enough vegetables. The military response was to add sugar to the few vegetables they had to feed the men. It was thought that this would make the food more palatable, but since sugar can suppress the immune system, the idea was far from helpful. However, both Washington and Napoléon's outfits offered apple cider vinegar and honey for soldiers to drink and, while they had little understanding of why it was healthful, it actually would have provided both vitamins and energy that would have helped keep the soldiers going. Food values and how the body processed food was not well understood.

This chapter outlines what people at this time understood about nutrition and the digestive and respiratory processes. An army surgeon and a gunshot victim led the way to better understanding of digestion. Antoine Lavoisier, the father of modern chemistry, made important contributions to understanding respiration.

WHAT THEY ATE

During the 1700s, the caliber of food was poor, particularly in urban areas. Meat rose in popularity, but it was difficult to transport in large quantities. Fresh fruit was also difficult to obtain, so the wealthy tended to be the primary consumers of anything fresh. There were also many misunderstandings about food preparation. The British thought anyone who ate uncooked fruit would get indigestion or even the plague. Another misunderstanding involved food and cleanliness. No one thought to wash what they ate. Fruit sold by vendors needed to look good, so one quick solution that was sometimes used was a little saliva from the fruit vendor; then he probably buffed the fruit on his largely unclean pants or shirt.

In the American colonies, life was not much better. The families who lived on farms worked hard. In the early settlements, poor families ate from trenchers filled from a common stew pot. The stews would have included pork, sweet corn, and cabbage or other vegetables and roots that were available.

As life became a little more refined, they would have eaten three meals. The caloric energy generated by a breakfast of cornmeal mush and molasses (washed down with cider or beer) would be used up quickly. By the 19th century, coffee, tea or chocolate were enjoyed by the fortunate.

The midday meal was generally the biggest meal of the day. While the affluent families would eat at home, stews were usually carried into the fields to feed the slaves and laborers. Supper, served at the end of the day, generally consisted of leftovers from dinner. Supper was generally more like a snack than a full meal, and, if times were difficult, it might have been gruel (a mixture made from boiling water with oats or cornmeal). Ale, cider, or some variety of beer were always served.

WHAT THEY KNEW ABOUT THE BODY

The idea that the body required constant nourishment went back hundreds of years, but the first controlled studies of the metabolic process in humans were undertaken by the 16th-century Italian physician Santorio Santorio (1561–1636). He saw the body as a machine and became interested in studying weight and its relation to food intake. Santorio created a balanced scale system that was big enough for him to sit in, and, over a 30-year period, he studied himself carefully. He described how he weighed himself before and after eating, sleeping, working, sex, fasting, drinking, and excreting.

While his findings ultimately did not have scientific value, his achievements were in the *empirical* methodology he used for the experiment. He was one of the first to pay such careful attention to gathering and evaluating data. (A better understanding of metabolism did not occur until the beginning of the 20th century when Eduard Buchner discovered enzymes. At this point it was possible to separate the study of the chemical reactions of metabolism from the biological study of cells, and this marked the beginning of biochemistry.)

THE DIGESTIVE PROCESS IN ACTION

In the early part of the 19th century, scientists realized that the stomach was key to the digestive process, but no one understood how it processed food—whether the food was ground up by the stomach, heated up so that it melted into a liquid form, or changed chemically. The military surgeon William Beaumont (1785–1853) was presented with a unique opportunity to study digestion, and he took full advantage of it.

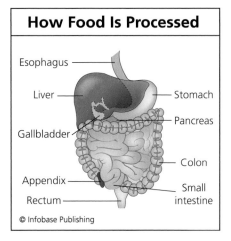

How Food Is Processed

Esophagus

Liver

Stomach

Pancreas

Gallbladder

Colon

Appendix

Rectum

Small intestine

© Infobase Publishing

William Beaumont's studies of how Alexis St. Martin processed food were early steps in understanding human digestion.

William Beaumont was born in Connecticut and trained to be a doctor. Because there were few medical schools in America, the most common way to study medicine involved studying under an established doctor, which Beaumont did with the local doctor in Lake Champlain. Beaumont was accepted as a physician by 1812, and he enlisted in the army as a surgeon's mate. He was assigned to a regiment in Plattsburgh, New York, where he took care of soldiers who were mainly suffering illnesses caused by the wet and windy weather. In April 1813, the regiment moved into battle, and the problems became more serious. In addition to battle wounds that often required amputations, the men also encountered more troublesome illnesses including dysentery, pleurisy, and pneumonia. Beaumont's treatment methods primarily involved wine, opium, mercury, and snakeroot. He used wood resin and turpentine for those suffering from rheumatism pain. Trephination was still used for pain relief as well, and Beaumont occasionally provided relief by cutting a small hole in the skull.

Later on, Beaumont was serving at Fort Mackinac, Michigan, when fate was to change two men's lives forever. A young French-Canadian voyageur (canoe paddler and trader) named Alexis St. Martin had stopped into the American Fur Company at Fort Mackinac for supplies when a musket discharged accidentally just

2.5 feet (0.76 m) from where St. Martin was standing. Beaumont, the fort doctor, was summoned right away, and he found that St. Martin had a hole bigger "than the size of the palm of a man's hand." In addition, part of the young man's lung was damaged and two ribs were broken. Beaumont did all he could to repair the wound, but the injury was so great that Beaumont felt St. Martin would be lucky to live 36 hours.

To everyone's amazement, St. Martin pulled through. The nature of the wound meant that he could no longer paddle canoes, so Beaumont hired him as a handyman to work at the fort. A year later, St. Martin was doing well, but the wound had still not completely closed. An opening into the stomach about 2.5 inches (6.35 cm) in circumference remained. Food and drink oozed out unless the area was bandaged.

BEAUMONT SEIZES AN OPPORTUNITY

For three years, the two men continued their separate lives at the fort; Beaumont tended to medical duties and St. Martin helped with whatever needed to be done. When Beaumont was transferred to Fort Niagara, he took St. Martin with him. At about this time, it occurred to Beaumont that he might be able to learn a lot about digestion by studying the French-Canadian. With St. Martin's agreement, Beaumont developed some experiments. One of them involved tying a silk string around different types of bite-sized morsels. Beaumont used various types of meat, stale bread, and cabbage, and he inserted the food directly into the hole, and then pulled it out via the string after various periods of time—one hour, two hours, and three hours. That day's experiments ended after five hours when St. Martin complained of stomach distress.

A few days later, Beaumont wanted to study digestion in and out of the stomach to see how gastric juices worked. (He did not know about the contributions of saliva.) Beaumont checked the temperature of the young man's stomach—it was 100°F (37.8°C). He also withdrew some of the gastric juice and put it in a test tube.

Keeping the test tube gastric juices at the same temperature as St. Martin's stomach, he introduced the same type of meat into both "test environments." He found that meat could be digested in the stomach in about two hours; the meat in gastric juice in the test tube took about 10 hours to digest. In September, St. Martin went back to Canada where he married and began raising what grew to be a large family.

Beaumont continued his army service, and, after stints in Green Bay, St. Louis, and Prairie du Chien, Wisconsin, Beaumont was reunited with St. Martin who agreed to return—for a fee—to continue the experiments. St. Martin and his family joined Beaumont in 1829, and the experiments continued. During this visit, Beaumont decided to observe "normal digestion." St. Martin would eat and then go back to work, and Beaumont would take samples from St. Martin's stomach at various times. This experiment showed Beaumont that milk coagulates before the digestive process, and vegetables take longer to digest than other foods. He also noted that if St. Martin was stressed, digestion took longer.

In 1832, Beaumont took a leave from the military and traveled with St. Martin to Washington. This time, Beaumont used oysters, sausage, mutton, and salted pork to test digestion. In 1833, Beaumont wrote about what he had learned, publishing *Experiments and Observations on the Gastric Juice and the Physiology of Digestion.*

The death of one of his children caused St. Martin to return to Canada, and, though the two men expected to get together again, St. Martin started asking for sums that exceeded what Beaumont could pay, and, as a result, the two men never worked together again.

Beaumont died before St. Martin; St. Martin lived to be 86, 58 years after the gunshot accident. St. Martin maintained a warm relationship with Beaumont's family until his own death in 1880. St. Martin's family felt St. Martin had suffered enough, and they did not want him to become a medical curiosity. They let his body decompose for several days and then buried him in the Catholic churchyard in a deep, unmarked grave and placed heavy rocks on top of the coffin to prevent anyone from performing an autopsy.

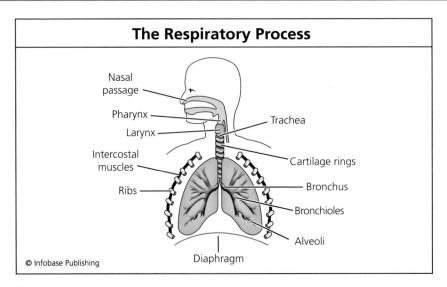

The Respiratory Process

Nasal passage

Pharynx

Larynx

Intercostal muscles

Ribs

Trachea

Cartilage rings

Bronchus

Bronchioles

Alveoli

Diaphragm

© Infobase Publishing

Through the work of scientists such as Antoine-Laurent de Lavoisier, physicians were beginning to understand more about the respiratory process.

Many years later, a committee finally persuaded one of St. Martin's granddaughters to disclose the location of the grave, and, in 1962, a plaque was placed on the church wall nearby, stating Alexis St. Martin's contribution to medical history.

THE EARLY WORK OF LOUIS PASTEUR (1822–1895)

Louis Pasteur was a giant among scientists, and his discoveries of germs and his work in vaccination will be discussed in a future volume of this series. However, early on, he made a great contribution that eventually would have a positive effect on public health by improving the quality of what people consumed.

Pasteur was born in 1822 in eastern France. He was not a particularly good student in elementary school, but one teacher saw possibilities and worked with him to teach him to take a very careful approach to his work. He went on to secondary school,

ANTOINE-LAURENT DE LAVOISIER (1743–1794):
The Father of Modern Chemistry

Antoine-Laurent de Lavoisier was one of the leaders in the amazing discoveries that were occurring in chemistry. In his work, Lavoisier developed an understanding of the chemical reactions of both combustion and respiration. This work resulted in the identification of oxygen, which was a vital discovery that could help with a better understanding of the workings of the body.

In 1777, Lavoisier identified that respiration involved the intake of oxygen and the exhalation of carbon dioxide. He went on to figure out ways to measure the oxygen intake for different activities. Further proof of Lavoisier's oxygen theory came when Lavoisier successfully decomposed water into two gases; he named them hydrogen and oxygen and later reformed them into water. As he continued his work, Lavoisier explored more about the passage of gases through the lungs and established that oxygen was indispensable for the human body.

Though unusual for the day, Lavoisier's wife, Marie-Anne Pierrette Paulze, became his colleague. She learned English so she could translate the work of English scientists for Lavoisier, and she developed skills in art and engraving and provided the illustrations for his books. She also left drawings that showed the devices with which he worked.

Though Lavoisier is considered the greatest chemist of his time, he was caught up in the government turmoil of the day. He was put to death (by guillotine) by the revolutionary government for being a member of the hated tax bureaucracy of the earlier regime.

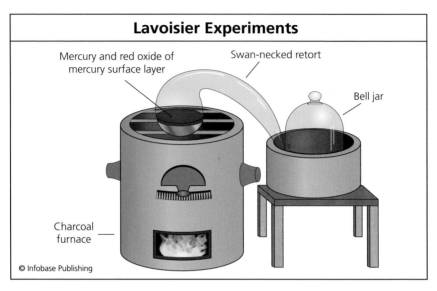

Lavoisier Experiments

Mercury and red oxide of mercury surface layer

Swan-necked retort

Bell jar

Charcoal furnace

© Infobase Publishing

This was the device Lavoisier used to disprove the theory that humans relied on phlogiston in order to live. The process involved heating mercury for 12 days and releasing it slowly. Once Lavoisier ascertained the phlogiston did not result from this process (as others said it would), it cleared the way for him to identify oxygen and carbon dioxide.

followed by the Ecole Normale in Paris, thinking he would train to be a teacher. Chemistry became the subject of his focus, and he graduated with a master of science in 1845. He began work toward a doctoral degree, and he thrived at the detailed work necessary in the laboratory. (The teacher who had encouraged careful and well-organized work habits is perhaps responsible for one of Pasteur's greatest qualities.) He went on to become professor of chemistry at the University of Strasbourg, and he married and started a family.

PRACTICAL SCIENCE

At the age of 32, Pasteur became part of a program where science faculty was expected to help apply their theoretical knowledge to work to solve the practical scientific problems of business and industry. Pasteur found this very exciting and spent two years establishing a faculty to work with him in applied science. His own research had to do with the process of fermentation—the process which is used to produce alcohol from sugar but which can also result in milk going sour. Chemists of the time could not explain why this was a good thing with wine but a bad thing with milk.

Pasteur proved that fermentation took place only when small living things called microbes were present. Pasteur discovered that spoilage organisms could be made inactive in wine by applying heat at temperatures just below its boiling point. The process was later applied to milk and remains an important part of keeping milk supplies safe. Pasteur's findings helped established a new branch of science—microbiology.

CONCLUSION

While there were still many unknowns about the respiratory and digestive processes, the 1700s and early 1800s were a time when scientists and physicians were beginning to put together some important pieces. No one could have dreamed of the opportunity given to Beaumont to learn about the inner workings of human digestion, and he and St. Martin contributed greatly to progress in this area. The study of respiration took a new leap forward with Lavoisier's work. From here, scientists could begin to study how oxidation takes place within the body—something they could not have learned without Lavoisier.

8

The Importance of Public Health

Toward the end of the Middle Ages, communities—particularly those in northern Italy—had encountered such problems with the spread of plague and other illnesses that many of the cities had established permanent boards of health that could establish quarantines, issue health passes, arrange for the burial of plague victims, and see that victims' homes were fumigated. Most boards worked closely with the local physicians who often advised them. Over time, some of the communities provided their boards with responsibilities for controlling the cleanliness of streets and marketplaces, in addition to maintaining adequate water supplies and sewage systems. Some cities placed the professional activities of physicians and surgeons and the monitoring of activities by beggars and prostitutes under the purview of the health boards as well. However, many of the boards of health ceased to be taken seriously during the 17th century when the plague's virulence lessened. At the time, towns did not seem to suffer from a lighter level of public health vigilance.

Not all cities in Western Europe had official health boards, but most implemented some measures that were helpful in controlling illnesses. Port cities regularly insisted on quarantine of sailors on newly arrived ships. Since miasma was still a popular theory of what made people sick, this led to some improvement

116

in sanitation as the population was motivated to get rid of odiferous things. Most communities tried to bury their dead quickly, and, while there was interest in getting rid of waste, few places had devised an efficient method for disposing of it. (Waste was frequently dumped into rivers and streams from which towns drew their freshwater.) What to do with the sick was always a problem. Towns frequently supported housing sick people with little money in "pest houses" as a way to halt contagion and get rid of a problem.

Johann Peter Frank (1745–1821), a leading clinician, medical educator, and hospital administrator, was one who recognized that public health was the key to solving many problems, and he dedicated his life to working toward creating governmental regulations and programs that protected the population against disease and promoted health. The actions that he advocated ranged from measures of personal hygiene and medical care to environmental regulation and social engineering. He was joined by other reformers who improved public health in their countries.

This chapter outlines the work of Frank and others like him who recognized a community problem and would not let go until it was solved. The work of John Snow, a British physician who made important contributions to the history of medicine as he unraveled the mystery of cholera, is explained.

EARLY AWARENESS

The period from 1750 until the mid–19th century was a time of unprecedented industrial, social, and political development. As the Industrial Revolution picked up steam and an ever-increasing number of people began to settle in the cities, the city governments were not prepared to handle the influx of so many people. Diseases like consumption, dysentery, smallpox, and typhus spread quickly through crowded communities. Many of the poor died from being undernourished, and the severe winters frequently led to illnesses from which those with little means did not recover.

The first sanitary commission in the United States was formed in 1861 to promote clean and healthy conditions in the Union army camps. They staffed some field hospitals and attempted to educate the military and government concerning cleanliness and healthy living.

Soon the death rates in urban areas began to exceed birth rates, and only the influx of people from the countryside kept the population growing.

The urban poor as a group also saw a decline in their life expectancy. In the industrial town of Manchester, England, in 1842, a member of the gentry could expect to live to age 38, but a factory worker's life expectancy was only 17 years! Outside the city, a craftsperson or laborer—the type of person who would have taken a factory job—could expect to live to 38 years. These same types of statistics were reflected in infant mortality: In the upper and middle classes, the death rate for babies was 76 per 1,000 births; for the unskilled laboring class, 153 deaths per 1,000 births was the norm. If poor children lived beyond toddlerhood, they generally suffered from issues related to poor nutrition. Rickets (softening of the bones) was particularly prevalent because of poor diets.

The historian Roy Porter, author of *The Greatest Benefit to Mankind: A Medical History of Humanity,* compares the industrial cities of that time with today's Third World shantytowns and refugee camps, with gross overcrowding, pollution of the water supply, and cesspools that frequently overflow, causing waste to run down the streets. The problems in these communities can be outlined by discussing two particular issues that plagued most cities: over-crowding and poor sanitation.

URBAN CROWDING

Housing for the majority of people who lived in the 19th century was incredibly bad. Many houses were poorly built, to the point of being unsafe. The rooms did not have lights or ventilation, and many had a dank or damp feel. Most people had to live in group housing, and, if a family did have their own space, they generally had only one bed that everyone slept in together. (It was called bundling.) Most houses had a fireplace that was used for cooking and for heat.

Between the dampness and the close quarters, the living environment encouraged the spread of diseases, and paying for medical care was unthinkable for most. Governments were not prepared to play a role in overseeing social welfare, so up until the first quarter of the 19th century, most forms of public medical assistance were provided by charitable organizations, idealistic doctors, and clergymen who simply volunteered to help out.

A LACK OF SANITATION

The streetscape and general town environment were shared by rich and poor, and sanitation was poor. The exposure to disease-carrying waste products became larger, as did the problem of unclean air. Noxious gases from burning coal and other types of industrial progress often caused a black or gray overlay to the air. In London, as in most cities in western Europe, very little was done to address these health concerns other than to force the more unsanitary industries such as leather tanning, glue-making, and candle-making out of the city into areas that were slightly less populated.

Eventually, outbreaks of large-scale infectious diseases began to force change. In the 1830s, typhus and cholera became rampant, and governments and local councils began to pay attention to the appalling conditions and the risks posed by contaminated wells, the lack of sewage systems, and people living in overcrowded housing. (See "John Snow and Cholera" on page 124)

Lack of Job Safety

No factory owner and few town administrators gave a thought to protecting workers from dangerous job conditions. Part of the problem was a lack of knowledge. The people of the time had no idea that working in a mine did damage to the lungs nor that the chemicals

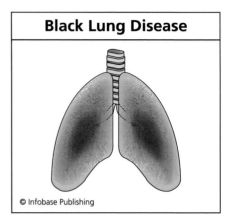

Black Lung Disease

© Infobase Publishing

This shows the type of lung damage that occurs to those who work in mines. Scientists were just beginning to realize that workplaces created unique hazards for workers. Very little was done about it at the time.

used in tanning posed a long-term health threat. As a result, no one paid much attention to creating safer working conditions. If someone was severely injured on the job, he was let go and expected to pay all medical costs himself. If he could not do so, then he likely received no care. The public sentiment of the time was that government should not interfere with employment practices or raise taxes in order to help the poor, as that was a violation of personal freedom. It took until the end of the 19th century before governments began to step in with laws and aid to protect workers.

JOHANN PETER FRANK (1745–1821): EARLY LEADER IN PUBLIC HEALTH

Johann Peter Frank is without a doubt one of the most influential figures in the early history of public health and community medicine. He was a physician who taught at several different universities and also worked as director of sanitation in Lombardy (1786) and as a sanitary officer to the Vienna hospitals (1795).

CHILDREN IN THE WORKPLACE

During the 18th and early 19th century, poor children were sent to work at a very young age. Some ran errands, swept roads, or sold flowers on the street. Many worked alongside their parents, sewing clothes or helping to make shoes that the family would sell. As the Industrial Revolution required more and more workers, many children began to work in factories, often running dangerous machinery. Hours were long and pay was poor.

Most poor children were in terrible health. They were often malnourished, and rickets (softening of the bones) was prevalent because of inadequate diets.

Sometimes children were sought out for jobs because being small was helpful in doing the work. Chimney sweeps loved having small children to go up into the chimneys to clean them. In factories, cotton-spinning machines were best operated by tiny fingers, and, because children learned quickly, they were put to work in these jobs. Factory operators often looked for children between the ages of six and 12 for this type of work.

The first effort to advocate for children came from charitable groups who organized missions where they provided employment—but it was thought, better employment—for children. While this may have kept the children from nefarious factory bosses who exploited them, it still prevented them from attending school or obtaining more helpful training.

Over time, governments began to put laws in place that were somewhat protective. In England, the Factory Act of 1833 proclaimed that children could not work until the age of nine and that children between the ages of nine and 13 could work only 48 hours a week. This was the first of several child labor laws to be enacted, but misuse of children continued into the 20th century.

Johann Peter Frank *(National Library of Medicine)*

At one point he was personal physician to Czar Alexander I (1805–08). Frank's work made him very concerned about public hygiene, and he undertook to devise codes of hygiene.

Early in his career he began working on a massive treatise, *System einer vollständigen medicinischen polizey* (A complete system of medical policy). This occupied him throughout his life, and, when it was published (1779 to 1827), it filled nine volumes. It was the first thorough treatise on all aspects of public health and hygiene, providing guidelines on an orderly method to keep communities clean. Frank's system dealt with water supply and sanitation, food safety, school health, sexual hygiene, maternal and child welfare, and regulation of aspects of public behavior. In addition, the treatise documented existing laws and proposed further regulations regarding conduct that affected people's health. He urged international regulation of health problems and advocated that one of the responsibilities of government was to protect the health of its citizens.

In his hospital work, Frank stressed the importance of keeping accurate statistical records for hospitals, and it may have been this system for maintaining health data that permitted Ignaz Semmelweis (1818–65) to demonstrate the connection between puerperal sepsis and unsanitary obstetrical practices.

In addition to his nine-volume *System,* Frank wrote a seven-volume textbook on internal medicine and made important clinical discoveries, including the distinction between different types of diabetes.

OTHER REFORMERS: BENTHAM, CHADWICK, AND SHATTUCK

In England, the social reformer Jeremy Bentham (1748–1832) was also pushing for a more humanitarian social philosophy. He believed that society should be organized for the greatest benefit of the greatest number of people (known as utilitarianism). He advocated for prison reform, various sanitary measures, and the establishment of a ministry of health and birth control.

The British government official Edwin Chadwick (1800–90) was a disciple of Bentham and worked to help Bentham's ideas become a reality. Chadwick had been secretary of England's Poor Law Commission, and he took the lead in advocating for trying to decrease the spread of disease among the poor, particularly the working poor. The resulting publication, *General Report on the Sanitary Condition of the Labouring Population of Great Britain* (1842), is considered one of the most important documents of modern public health. His report included figures to show that in 1839 for every person who died of old age or violence, eight died of specific diseases. (These statistics help explain why during the second and third decades of the 19th century nearly one infant in three in England failed to reach the age of five.)

In his work, Chadwick documented the life expectancy of various social classes, the status of housing of the working population, the lack of adequate supplies of water, and the existence of poor sewage disposal. He also noted the unhygienic circumstances of most workplaces, the economic impact of unsanitary conditions, and the evidence for the beneficial health effects of preventive measures.

Chadwick's report was widely circulated and carefully considered, and over time legislation began to be introduced that provided for better sewage, adequate clean water supplies, regular refuse removal, and ventilation for homes and in factories. Chadwick also fought for laws that might help reduce workplace injuries.

In the United States, the reformer Lemuel Shattuck observed the impact that Chadwick was having, and Shattuck put in place the mechanism for a similar survey with recommendations based on what was learned. *The Report of a General Plan for the Promotion of Public and Personal Health* (1850) put forward 50 recommendations and a model for state public health laws.

In England and America, these reports began to have definite impacts on the governments, establishing a framework for an improvement in the field of public health.

JOHN SNOW AND CHOLERA

As previously noted, a more mobile population led to a greater spread of illness as diseases from other countries arrived with explorers, travelers, and traders as they returned home. In 1816, cholera—an acute disease that is characterized by violent stomach-related problems—began to spread rapidly from India to the ports of the Philippines, China, Japan, the Persian Gulf, and then north toward the Ottoman and Russian Empires, killing thousands of people. The first outbreak of Asiatic cholera, as it is sometimes referred to, in Britain was at Sunderland on the Durham coast during the autumn of 1831. The disease traveled north to Scotland and south toward London. By the end of that outbreak, 52,000 lives were lost. Then in 1832, London experienced another outbreak that killed 7,000 people.

In *The Healthy Body and Victorian Culture,* Bruce Haley quotes local doctors:

John Snow *(National Library of Medicine)*

. . . cholera was something outlandish, unknown, monstrous; its tremendous rav-

London: How Cholera Spread

This was the type of map John Snow used to identify the location of those who died from cholera. Using that information, he was able to trace cholera back to the pumps that were supplied with water from a company that took the liquid from a polluted part of the Thames.

ages, so long foreseen and feared, so little to be explained . . . its apparent defiance of all the known and conventional precautions against the spread of epidemic disease . . . recalled the memory of the great epidemics of the Middle Ages.

Symptoms of cholera are nausea and dizziness that lead to violent vomiting and diarrhea. Extreme muscle cramps follow with an insatiable desire for water, followed by a sinking stage when the pulse rate drops and lethargy sets in. Near death, the patient displays the classic cholera look, which features puckered blue lips in a face that becomes very skeletonlike.

John Snow (1813–58), a British physician, was particularly puzzled by cholera, but he also was exploring a completely different theory about the spread of illness. Snow believed that disease could be carried by contaminated food or water, and in 1849 he published a small pamphlet "On the Mode of Communication of Cholera." Most professionals still believed that disease was transmitted by contaminated vapors. While a few scientists took note of Snow's idea that the Cholera poison was being spread by contaminated food or water, he was largely ignored.

Then in 1854, England experienced a terrible outbreak of cholera, and Snow set about investigating the epidemic and mapping out the locations of those who were dying of the illness. At the time, the London public received water from two water companies. One took water from the Thames, upstream of the city; the second company also took water from the Thames, but their source was downstream of the city. The cases of cholera seemed to be clustered around the pumps and wells that collected their water from the downstream source. Snow also noted that one particular water pump seemed to be in the center of an extraordinarily high outbreak of the disease. According to his map, there were up to 500 deaths from cholera during a 10-day period near a pump at Cambridge and Broad Streets.

As a first step, Snow suggested that public officials remove the pump handle from the Broad Street pump, and, to everyone's great surprise and relief, the number of cases in the area near the Broad Street pump began to drop quickly. While later scientists would verify Snow's suspicion that the causative factor for the spread of cholera was an unknown agent in the water, the decrease in cases after the change of the pump handle likely had to do with the fact that people did not wash their hands often. The pump handle must have been highly contaminated.

As Snow pushed for cleaner water, other scientists were working to explain what Snow suspected. In 1883, the chemist Robert

This sketch, titled *Death's Dispensary,* was drawn by George Pinwell in 1866, about the time John Snow published his studies that showed the source of cholera to be the water supply. *(Public Health Image Library, Centers for Disease Control)*

Koch identified *Vibreo cholerae.* But even before Koch's contribution, Snow's work was absolutely vital to establishing a new precedent for how to look for the cause of disease. As a result of this work as well as other contributions to medicine, Snow is often referred to as the father of *epidemiology.*

CONCLUSION

During the mid-19th century, Europe was experiencing a time of great unrest. Revolutions in France, Germany, Hungary, Italy, and the Habsburg Austrian Empire created harsh living conditions for most of the population, which eventually brought greater focus to the issues involved in public health. The Irish Potato Famine (1845–51) also contributed to additional awareness of the need for reform. When Ireland's crops failed, it caused the deaths of 1 million people, with another 1 million leaving the country in coffin ships to try and escape the great hunger. While few were particularly concerned about the poor, leading citizens and governments began to realize that something needed to be done to improve life for everyone. Slowly, new public health laws began to be put in place.

CHRONOLOGY

1630s	Plants such as Peruvian bark, tobacco, and cinchona begin to be imported from the New World and used as medicines.
1683	Leeuwenhoek sees "little aminalcules."
1692	Salem witch trials
1700s	Men begin to take on role of midwives, called accoucheurs.
1708	*Institutiones medicae* by Herman Boerhaave published
1733	Forceps begin to be used by people other than the Chamberlens.
1740	Sir John Pringle identifies typhus.
1744–1906	Patent medicines are extraordinarily popular.
1747	James Lind runs clinical trial and proves that citrus fruits prevent scurvy.
1752	John Pringle publishes the first English text on military medicine, *Observations on the Diseases of the Military.*
	Britain passes the Murder Act, which somewhat eases availability of bodies for dissection.
1763	Smallpox-infected blankets were distributed by the British to Native Americans, starting an epidemic; there is disagreement about whether it was intentional.
1770–93	Surgeon John Hunter is at his peak.
1774	William Hunter publishes his seminal work on pregnancy.
1777	Lavoisier identifies that respiration involves intake of oxygen.
1788	People riot at the hospital over doctors' use of bodies for study.

1790–1810	Leeches are so popular that Europe has to import them.
1790s	Mesmerism becomes the rage.
	Samuel Hahnemann objects to bloodletting and begins teaching homeopathy.
1793	Philadelphia experiences devastating yellow fever epidemic.
1796	Edward Jenner develops a smallpox vaccination.
1797	Larrey develops concept of flying ambulance corps.
	Larrey implements the use of triage to prioritize treatment of the wounded.
1799	The death of George Washington, in spite of—or because of—medical care
	Benjamin Rush wins libel suit against writer William Cobbett, who has attacked him in print for Rush's copious "medicinal" bloodletting.
1800	Humphry Davy identifies nitrous oxide.
1806	First American medical licensing law passed; abolished 1844
1811	Elemental iodine discovered
1816	René Laennec invents the stethoscope.
1820–40s	Phrenology becomes popular.
1825	Homeopathy becomes popular in America.
1832	The Anatomy Act passed by Britain, further easing availability of bodies.
1841	Dr. James Braid develops hypnosis.
1842	First surgical operation using anesthesia
1847	American Medical Association (AMA) formed
1850s	William Beaumont begins experiments to understand digestion.
1860s	Lister experiments with ways to create sterile environment.

1862	Pasteur refines what becomes known as pasteurization.
1899–1901	Walter Reed heads a commission that finally determines that yellow fever is spread by mosquitoes; this provides a way to diminish the contagion.
1906	United States passes Pure Food and Drug Act.
1978	Last case of smallpox; person died after accidental exposure.
1980	Leeches begin being used again in surgery.

accoucheur   one that assists at a birth; an obstetrician

acute   being, providing, or requiring short-term medical care (as for a serious illness or traumatic injury)

blister   an elevation of the epidermis containing watery liquid

caliper   any of various measuring instruments having two usually adjustable arms, legs, or jaws used especially to measure diameters or thickness—usually in pl.

calomel   a white tasteless compound, formerly used in medicine as a purgative—called also mercury chloride

cantharides   a preparation of dried beetles (as Spanish flies) used in medicine as a counterirritant and formerly as an aphrodisiac

cartouche   a box for cartridges

coagulate   to cause to become viscous or thickened into a coherent mass; to clot

deductive reasoning   of, relating to, or provable by deduction

edema   an abnormal infiltration and excess accumulation of serous fluid in connective tissue or in a serous cavity—called also dropsy

empirical   originating in or based on observation or experience

epidemiology   a branch of medical science that deals with the incidence, distribution, and control of disease in a population

farrier   a person who shoes horses

felony   a grave crime formally differing from a misdemeanor under English common law by involving forfeiture in addition to any other punishment

frigate   a light boat propelled by sails

gout   a metabolic disease marked by painful inflammation of the joints and usually an excessive amount of uric acid in the blood

homeopathy   a system of medical practice that treats a disease especially by the administration of minute doses of a remedy that would in healthy people produce systems similar to those of the disease

hypotension   abnormally low blood pressure

inductive reasoning   of, relating to, or employing mathematical or logical induction

jalap   a dried tuberous root of a Mexican plant (*Ipomoea purge syn. Exogonium purge*) of the morning-glory family; also, a powdered purgative drug prepared from it that contains resinous glycosides

jaundice   yellowish pigmentation of the skin, tissues, and body fluids caused by the disposition of bile pigments

laudanum   any of various preparations of opium

leeches   any of numerous carnivorous or bloodsucking usually freshwater annelid worms that have typically a flattened segmented body with a sucker at each end

lying-in hospital   hospital for childbirth

miasma   a vaporous exhalation formerly believed to cause disease; also, a heavy vaporous emanation or atmosphere

misdemeanor   a crime less serious than a felony

nostrum   a medicine of secret composition recommended by its prepared but usually without scientific proof of its effectiveness

panacea   a remedy for all ills or difficulties, a cure-all

patent medicine   a packaged nonprescription drug that is protected by a trademark and whose contents are incompletely disclosed; also, any drug that is a proprietary

phrenology   a study of the conformation of the skull based on the belief that it is indicative of mental faculties and character

plethora   a bodily condition characterized by an excess of blood and marked by turgescence and a florid completion

potter's field   a public burial place for paupers, unknown persons, and criminals

poultice   a soft, usually heated, and sometimes medicated mass spread on cloth and applied to sores or other lesions

quarantine   a restraint upon the activities or communication of people or the transport of goods designed to prevent the spread of disease or pests

quinsy   an abscess in the tissue around a tonsil usually resulting from bacterial infection and often accompanied by pain and fever

resurrectionist body snatcher; people who robbed graves for bodies to sell to medical schools for dissections, i.e., William Burke and William Hare

shill to act as a promoter

styptic tending to check bleeding

tartar a poisonous efflorescent crystalline salt of sweetish metallic taste formerly used in medicine as an emetic and expectorant

vaccine a preparation of killed microorganisms, living attenuated organisms, or living fully virulent organisms that is administered to produce or artificially increase immunity to a particular disease

variolation the obsolete process of inoculating a susceptible person with material taken from a vesicle of a person who has smallpox

vasodilator dilation or relaxation of the blood vessels

venesection cutting of a vein

FURTHER RESOURCES

ABOUT SCIENCE AND HISTORY

Diamond, Jared. *Guns, Germs, and Steel: The Fates of Human Societies.* New York: W. W. Norton and Company, 1999. Diamond places in context the development of human society, which is vital to understanding the development of medicine.

Dubus, Allen G. *Man and Nature in the Renaissance.* Cambridge: Cambridge University Press, 1978. Includes quotes from Vesalius that were very helpful in understanding his work.

Hazen, Robert M., and James Trefil. *Science Matters: Achieving Scientific Literacy.* New York: Doubleday, 1991. A clear and readable overview of scientific principles and how they apply in today's world, including the world of medicine.

Internet History of Science Sourcebook. Available online. URL: http://www.fordham.edu/halsall/science/sciencebook.html. Accessed July 9, 2008. A rich resource of links related to every era of science history, broken down by disciplines, and exploring philosophical and ethical issues relevant to science and science history.

Lindberg, David C. *The Beginnings of Western Science, Second Edition.* Chicago: University of Chicago Press, 2007. A helpful explanation of the beginning of science and scientific thought. Though the emphasis is on science in general, there is a chapter on Greek and Roman medicine as well as medicine in medieval times.

Roberts, J. M. *A Short History of the World.* Oxford: Oxford University Press, 1993. This helps place medical developments in context with world events.

Silver, Brian L. *The Ascent of Science.* New York: Oxford University Press, 1998. A sweeping overview of the history of science from the Renaissance to the present.

Spangenburg, Ray, and Diane Kit Moser. *The Birth of Science: Ancient Times to 1699.* Rev. ed. New York: Facts On File, 2004. A highly readable book with key chapters on some of the most significant developments in medicine.

ABOUT THE HISTORY OF MEDICINE

Ackerknecht, Erwin H., M.D. *A Short History of Medicine, Revised Edition.* Baltimore, Md.: Johns Hopkins University, 1968. While there have been many new discoveries since Ackerknecht last updated this book, his contributions are still important as they help the modern researcher better understand when certain discoveries were made and how viewpoints have changed over time.

Bell Jr., Whitfield J. "Doctor's Riot, New York, 1788." *Bulletin of the New York Academy of Medicine* 47, no. 12 (December 1971): 1,501–1,503. This article contains a firsthand account of the riot that took place over physicians dissecting cadavers.

Bishop, W. J. *The Early History of Surgery.* London: The Scientific Book Guild, 1960. This book is dated but helpful on the history of surgery.

Buchan, William. *Domestic Medicine, Second Edition.* London: Royal Society, 1785. Available online. URL: http://www.american revolution.org/medicine.html. Accessed January 10, 2009. Provides a contemporary account of the medical beliefs of the late 1700s.

Carlson, Laurie Winn. *A Fever in Salem.* Chicago, Ill.: Ivan R. Dee Publishers, 1999. A new interpretation of what might have affected the girls who were thought to be under a witch's spell.

Chambers, Robert, ed. *Biographical Dictionary of Eminent Scotsmen.* Glasgow, Edinburgh, and London: Blackie and Son, 1856. Available online. URL: http://www.electricscotland.com/history/other/hunter_william.htm. Accessed January 10, 2009. This resource provides excellent information on the Scottish physicians, William and John Hunter.

Clendening, Logan, ed. *Source Book of Medical History.* New York: Dover Publications, 1942. Clendening has collected excerpts from medical writings from as early as the time of the Egyptian papyri, making this a very valuable reference work.

Dary, David. *Frontier Medicine: From the Atlantic to the Pacific 1492–1941.* New York: Knopf, 2008. This is a brand new book that has been very well reviewed; Dary is a western historian, and he outlines the medical practices in the United States from 1492 on.

Davies, Gill, ed. *Timetables of Medicine.* New York: Black Dog & Leventhal, 2000. An easy-to-assess chart/time line of medicine with overviews of each period and sidebars on key people and developments in medicine.

Dawson, Ian. *The History of Medicine: Renaissance Medicine.* New York: Enchanted Lion Books, 2005. A heavily illustrated short book to introduce young people to what medicine was like during medieval times. Dawson is British so there is additional detail about the development of medicine in Britain.

Dittrick Medical History Center at Case Western Reserve. Available online. URL: http://www.cwru.edu/artsci/dittrick/site2/. Accessed October 31, 2008. This Web site is a helpful resource to link to medical museum Web sites.

Duffin, Jacalyn. *History of Medicine.* Toronto, Canada: University of Toronto Press, 1999. Though the book is written by only one author (a professor), each chapter focuses on the history of a single aspect of medicine, such as surgery or pharmacology. It is a helpful reference book.

Dunn, Peter M. "The Chamberlen Family (1560–1728) and Obstetric Forceps." *Archives of Disease in Childhood.* Fetal and Neonatal Edition 81, no. 3 (November 1998): F232–F234. This is an enlightening perspective on the Chamberlen family and why they maintained their secret.

Fenn, Elizabeth Anne. *Pox Americana: The Great Smallpox Epidemic of 1775–82.* New York: Hill & Wang, 2001. This is a scholarly book that describes the devastating impact of smallpox in North America.

Haeger, Knut. *The Illustrated History of Surgery.* Gothenburg: AB Nordbok, 1988. This is an academic book that is very helpful in understanding early surgery.

Haley, Bruce. *The Healthy Body and Victorian Culture.* Cambridge, Mass.: Harvard University Press, 1978. Haley's book provides insightful comments about how the Victorians felt about health care and taking care of themselves.

Kennedy, Michael T., M.D., FACS. *A Brief History of Disease, Science, and Medicine.* Mission Viejo, Calif.: Asklepiad Press, 2004. Michael Kennedy was a vascular surgeon and now teaches first and second year medical students an introduction to clinical medicine course at the University of Southern California. The book started as a series of his lectures but he has woven the material together to offer a cohesive overview of medicine.

Loudon, Irvine, ed. *Western Medicine: An Illustrated History.* Oxford: Oxford University Press, 1997. A variety of experts contribute chapters to this book that covers medicine from Hippocrates through the 20th century.

Magner, Lois N. *A History of Medicine.* Boca Raton, Fla.: Taylor & Francis Group, 2005. An excellent overview of the world of medicine from paleopathology to microbiology.

Ortiz, Jose P. "The Revolutionary Flying Ambulances of Napoleon's Surgeon." *U.S. Army Medical Department Journal* (October–December 1998): 17–25. Larrey's contributions to the military are outlined, and Ortiz refers to many primary sources in this article.

Porter, Roy, ed. *The Cambridge Illustrated History of Medicine.* Cambridge: Cambridge University Press, 2001. In essays written by experts in the field, this illustrated history traces the evolution of medicine from the contributions made by early Greek physicians through the Renaissance, Scientific Revolution, and 19th and 20th centuries up to current advances. Sidebars cover parallel social or political events and certain diseases.

———. *The Greatest Benefit to Mankind: A Medical History of Humanity.* New York: W. W. Norton Company, 1997. Over his lifetime, Porter wrote a great amount about the history of medicine, and this book is a valuable and readable detailed description of the history of medicine.

Rosen, George. *A History of Public Health, Expanded Edition.* Baltimore, Md.: Johns Hopkins University Press, 1993. While serious public health programs did not get underway until the 19th century, Rosen begins with some of the successes and failures of much earlier times.

Rush, Benjamin, M.D. *The Autobiography of Benjamin Rush: His "Travels Through Life" Together with His Commonplace Book for 1789–1813.* Reprint. Westport, Conn.: Greenwood Press, 1970. This provides Benjamin's Rush's perspective on medicine of his day.

Selwyn, S. "Sir John Pringle: Hospital Reformer, Moral Philosopher and Pioneer of Antiseptics." *Medical History* (July 10, 1966): 266–274. This provided an enlightening portrait of John Pringle.

Simmons, John Galbraith. *Doctors & Discoveries.* Boston: Houghton Mifflin, 2002. This book focuses on the personalities behind the discoveries and adds a human dimension to the history of medicine.

Toledo-Pereyra, Luis H. *A History of American Medicine from the Colonial Period to the Early Twentieth Century.* Lewiston, N.Y.: Edwin Mellen, 2006. This is an academic book that provides very valuable information about colonial America.

United States National Library, National Institutes of Health. Available online. URL: http://www.nlm.nih.gov/hmd/. Accessed July 10, 2008. A reliable resource for online information pertaining to the history of medicine.

Vadakan, Vibul V., M.D., FAAP. "The Asphyxiating and Exsanguinating Death of President George Washington." *The Permanente Journal* 8, no. 2 (Spring 2004): 76–79. Vadakan takes a clinical look at the treatment of George Washington's last illness.

OTHER RESOURCES

Collins, Gail. *America's Women 400 Years of Dolls, Drudges, Helpmates, and Heroines.* New York: William Morrow, 2003. Collins's book contains some very interesting stories about women and their roles in health care during the early days of America.

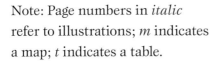

INDEX

Note: Page numbers in *italic* refer to illustrations; *m* indicates a map; *t* indicates a table.